T0271338

A Journey into the
World of Exponential Functions

Gautam Bandyopadhyay

CRC Press
Taylor & Francis Group
Boca Raton   London   New York

CRC Press is an imprint of the
Taylor & Francis Group, an **informa** business

Levant Books
India

First published 2023
by CRC Press
4 Park Square, Milton Park, Abingdon, Oxon, OX14 4RN

and by CRC Press
6000 Broken Sound Parkway NW, Suite 300, Boca Raton, FL 33487-2742

© 2023 Gautam Bandyopadhyay and Levant Books

CRC Press is an imprint of Informa UK Limited

The right of Gautam Bandyopadhyay to be identified as author of this work has been asserted in accordance with sections 77 and 78 of the Copyright, Designs and Patents Act 1988.

Trademark notice: Product or corporate names may be trademarks or registered trademarks, and are used only for identification and explanation without intent to infringe.

Print edition not for sale in South Asia (India, Sri Lanka, Nepal, Bangladesh, Pakistan or Bhutan).

ISBN: 9781032526867 (hbk)
ISBN: 9781032526881 (pbk)
ISBN: 9781003407850 (ebk)

DOI: 10.4324/9781003407850

Typeset in Knuth Computer Modern
by Levant Books

LEVANT

Dedicated in loving memory of my teacher Late Professor Basanta Kumar Samanta who taught us Calculus in the most enchanting way in the erstwhile Bengal Engineering College.

"In general, when telling a mathematical story, there are various goals such as elegance, rigour, practicality, generality and understandability. Sometimes these goals conflict, and we have to compromise. Sometimes developing a subject in the most logically concise way does not make for easy reading. As with any other subject, learning mathematics from multiple perspectives leads to deeper and more critical understanding."

Calculus: Module 14 Exponential and logarithmic functions
Australian Mathematical Science Institute

"Mathematics as an expression of the human mind reflects the active will, the contemplative reason, and desire for aesthetic perfection. Its basic elements are logic and intuition, analysis and construction, generality and individuality. Though different traditions may emphasize different aspects, it is only the interplay of these antithetic forces and the struggle for their synthesis that constitute the life, usefulness, and supreme value of mathematical science."

Richard Courant

Preface

This book is primarily meant for engineering students and others who need to apply mathematics for solving practical problems. This is not a conventional mathematics book which emphasizes on mathematical rigour while dealing with different mathematical topics.

Apart from routine calculations, a large section of educated people prefer to distance themselves from mathematics. The particular branch of mathematics that appeals to a certain individual, of course depends upon the person: for people whose interaction with the subject has ceased after college, this area of preference is nominal. However, even for people pursuing higher studies in exact sciences and engineering, the process of solving a problem is mechanical — a mere blind way of doing what certain rules tell them to do. Certain great mathematicians have laid down these rules and therefore they CANNOT be wrong. It is this, more than anything else, that makes most of us overlook the real spirit of mathematics. Like every other achievement, none of these mathematical procedures were created overnight. It took years (sometimes centuries) and acumen of several outstanding mathematicians to unravel these mysteries. Text books on mathematics play a great role in disseminating knowledge. However, they may not always enable a student to appreciate that knowledge since he/she remains unaware of the root of a novel theory and its evolution.

The number e, the function e^x (in its various forms), the function $\ln(x)$, different hyperbolic functions; $\cosh(x)$, $\sinh(x)$, etc. and complex angles (where real part is a hyperbolic angle and the imaginary part is a circular angle or vice versa) make frequent appearances in science and engineering text books. For most of us, our first encounter with these symbols was not very pleasant; we failed to appreciate the significance of these symbols. With time, however, we get used to these symbols though a sense of discomfort lies hidden in our minds. When I joined academics later, during my free time I made inquiries in different books and papers to alleviate my discomfort in this regard. Sometimes getting a cue from certain published literature I made some investigations myself. This book is an outcome of all these activities and I basically want to share my understanding of these mathematical concepts with the engineering fraternity at large. However, other users of these mathematical topics may also find some interest in this book.

Numbers play a central role in mathematics. Perhaps it started with simple ideas related to counting different objects. However, it has afterwards undergone a long process of evolution. Numbers like e, e^2, $\ln 3$ and $\sinh 1.5$ are transcendental numbers which were introduced at a later stage of development of mathematics. This book also deals with imaginary numbers and complex numbers (which are shrouded with mystery in the general mind) quite extensively. As such, Chapter 1 briefly discusses the evolution of numbers in mathematics.

Most of the mathematical topics start with simple ideas at the beginning; with time, however, these simple ideas get refined and enriched. Very often they accommodate other concepts and merge with them. As such, afterwards they may look very different. Mathematician Otto Toeplitz suggested that the genetic approach of teaching mathematics enables a person to grasp and appreciate complex mathematical ideas in an effective way. "Follow the genetic course, which is the way man has gone in the understanding of mathematics, and you will see that humanity did ascend gradually from the simple to the complex. Occasional explosive great development can usually be taken as indicators of preceding methodical progress. Didactical methods can thus benefit immeasurably from the study of history". Keeping this in mind effort has been made to trace the main line of development of the aforementioned mathematical concepts.

It is important to mention that unlike mathematicians, engineers and physicists want to visualize mathematical concepts and to learn an abstract concept through its concrete use. Aram Boyajian, a renowned electrical engineer of General Electric Company wrote "In as much as problems can be solved mathematically without any visualization, the mathematician discards visual interpretations and bases his conception on formulas. However, the engineering type of mind finds it both difficult and distasteful to be dependent on symbolic definitions, and craves for visualization. He feels that if physical problems lead to 'imaginary' or 'complex' angles, these angles must stand for some concrete physical facts and must therefore be capable of a physical interpretation." Being an electrical engineer I also look at the subject of mathematics from a similar angle and tried to explain the mathematical concepts following these guidelines.

Only a few stalwarts like Joseph Fourier (introduced Fourier Analysis to solve heat flow problem), Charles Proteus Steinmetz (introduced complex number based symbolic technique to study steady state behaviour of a.c. circuits) and Oliver Heaviside (introduced Operational Calculus based algebraic techniques to study transient behaviour of a.c. circuits) are gifted with the talent to frame suitable theoretical tools based on abstract mathematical ideas for the purpose of solving practical scientific and engineering problems. A disciplined mind with sound reasoning and expertise in a specific scientific area can only visualise the invisible connection between an abstract mathematical concept and the processes occurring in that area. But by the same logic only a few of us can write a novel or compose a song. However, most of us enjoy a good novel or a sweet song. The author of this book feels that by going back to the roots of the refined conceptions of many mathematical ideas it is possible to stimulate our minds. So the lifeless symbols after long mechanical use will be reborn with fresh vitality.

Professor Ananda Mohan Ghose, my teacher in erstwhile Bengal Engineering College, kindly read the first draft of the second chapter "On e & e^x" and offered valuable advice to me.

Professor Siddhartha Sen who has recently retired from IIT Kharagpur and Professor Soumitro Banerjee of IISER, Kolkata have kindly gone through the book and offered their valuable opinions to improve the quality of the book.

I am grateful to my wife Ranjana and daughters Rubenka and Riyanka for the love, support and encouragement throughout the time of writing this book.

I thankfully acknowledge the support extended by my colleagues Professor Narayan Chandra Maity, Professor Arabinda Roy, Professor Amit Das, Professor Prasid Syam, Professor Debjani Ganguli, Professor Aparajita Sengupta, Professor Aparna (Dey) Ghosh, Professor

Mainak Sengupta, Professor Amit Roy Chowdhuri and Professor Abhik Mukherjee in a variety of ways to prepare this book.

I am thankful to my student Sri Gourab Banerjee of IIEST, Shibpur who has drawn most of the diagrams of this book in spite of his busy academic schedule. I also thank Sri Pinak Pani Banerjee, a Project Associate in the Department of Electrical Engineering IIEST, Shibpur in this regard.

Shibpur, Howrah–711103   Gautam Bandyopadhyay
India   gautam.b.1955@gmail.com
October, 2022

Contents

Chapter 1

On Different Types of Numbers

"The child comes into a social environment which is possessed of a highly perfect number system but the child feels no need of it and he is incapable of abstract thinking that is necessary for the intelligent use of this system. To the child the number system in itself a body of very complicated experiences · · · The deliberate effort of society is to give pupils in a few years and in the most highly perfected form, that which the race strove during long centuries."

Judd, C. H.

1.1 The beginning of the journey

We will start from our childhood days (or perhaps the infancy of the human race) when mathematics was unknown to us. Even then we could distinguish between one bird and two birds or between two birds and five birds. We could also distinguish five men from three men or absence of any man from the presence of two men. We need not use mathematics to understand these distinctions. All of us have some intuitive feeling to make distinctions in the multitudes present in the physical world. These feelings remain the same irrespective of the entity we consider — be it birds or men or marbles or anything else for that matter. We find something similar/common in three dogs, three chairs and three bananas; we call it the number three. We have the inherent capacity to construct an abstract world of quantities and make distinctions in the multitudes of that physically non-existent world. In this world other features viz. the features which make a dog a dog or a chair a chair or a banana a banana — play no role at all. *This ability of constructing an abstract world inherent in all of us is the cornerstone of the subject of mathematics.*

Mathematics sets off its journey when we started writing symbols to represent our intuitive feelings about our quantitative world. To represent a single quantity we use the symbol '1', to represent two quantities — the symbol '2', and so on. The use of the symbol '0' to represent the absence of any quantity was conceived later since apparently it seems that absence of anything does not require any representation. In the first phase of this act of symbolisation we move from real world to the realm of symbols. Sometimes we move backwards from the symbolic world to the real physical world. In this way, while shuttling back and forth from the symbolic world to the physical world, we find a particular symbol becomes so intimate with its physical counterpart that they lose their individual identities by forming a single entity — unique by itself.

The job of symbolisation, however, did not remain so simple an affair. In the physical world, there may be a large number of entities (without any end) and if we keep on using a large number of unique symbols (without any end) for each of them we find ourselves moving through a blind alley. Long ago our ancestors, the Indian mathematicians, devised a very effective technique to solve this riddle. To illustrate this technique let us imagine that we have with us a big heap of peanuts (p.n.). As per this technique we need to use the ten symbols shown in Table 1.1.

For ten peanuts we may take a packet and put them inside it. So we have 1 packet and 0 loose peanut. We use two symbols 10 to represent ten peanuts — the first one for the number of packets and second one for the loose peanuts. Similarly for eleven peanuts we put ten peanuts in a packet leaving one loose peanut outside. So for eleven peanuts we use two symbols 11, the first 1 indicating the number of

Table 1.1: Single Digit Numbers

No	one	two	three	four	five	six	seven	eight	nine
p.n.	p.n.	p.n.s	p.n.s	p.n.s	p.n.s	p.n.s	p.n.s	p.n.s	p.n.s
0	1	2	3	4	5	6	7	8	9

packets and the second 1 indicating the number of loose peanut left outside the packet. We have shown in Table 1.2 the way we use two symbols to represent larger numbers.

Table 1.2: Two Digit Numbers

Ten p.n.s	Eleven p.n.s	..	..	..	..	Eighteen p.n.s	Nineteen p.n.s
10	11	..	..	..	..	18	19
Twenty p.n.s	Twenty one p.n.s	..	..	..	..	Twenty eight p.n.s	Twenty nine p.n.s
20	21	..	..	..	..	28	29
..	..	..	..	..	..	..	..
..	..	..	..	..	..	..	..
..	..	..	..	..	..	..	..
..	..	..	..	..	..	..	..
..	..	..	..	..	..	..	..
..	..	..	..	..	..	..	..
Ninety p.n.s	Ninety one p.n.s	..	..	..	..	Ninety eight p.n.s	Ninety nine p.n.s
90	91	..	..	..	..	98	99

When we have one hundred peanuts we may think it as 99 peanuts and one more. So we put 9 loose peanuts and the new peanut in one more packet. So we have ten packets with no loose peanut left outside. We now put ten packets inside one box with no packet and no loose peanut left outside. So for hundred peanuts we use three symbols 100 side by side where the leading 1 indicates the number of boxes (each box contains ten packets or hundred peanuts), next 0 indicates no packet is left out and next 0 indicates no loose peanut is left outside. We have shown in Table 1.3 the way we use three symbols to represent even larger numbers.

In this way we can keep on choosing larger and larger containers in our imagination and avoid the use of more than ten symbols $(0, 1, 2, \cdots, 8, 9)$ for each type of container. So, when we need to use ten boxes we put them inside a carton (an even larger container). We may now keep on using the symbols 0, 1, 2, $\cdots$, 8, 9 for the cartons to represent even larger number of peanuts.

Table 1.3: Three Digit Numbers

One hundred p.n.s	One hundred one p.n.s	..	..	One hundred nine p.n.s
100	101	..	..	109
One hundred ten p.n.s	One hundred eleven p.n.s	..	..	One hundred nineteen p.n.s
110	111	..	..	119
..	..	..	..	..
..	..	..	..	..
..	..	..	..	..
Six hundred p.n.s	Six hundred one p.n.s	..	..	Six hundred nine p.n.s
600	601	..	..	609
..	..	..	..	..
..	..	..	..	..
..	..	..	..	..
Nine hundred Ninety p.n.s	Nine hundred Ninety one p.n.s	..	..	Nine hundred Ninety nine p.n.s
990	991	..	..	999

Example 1:

A **packet** contains **10** peanuts.

A **box** contains 10 packets or **100** peanuts.

A **carton** contains 10 boxes or 100 packets or **1000** peanuts.

Peanuts	Thousand Carton	Hundred Box	Ten Packet	One
	7	5	4	6

Looking critically at this technique (Hindu positional system) the following inferences can be made.

1. Each of the ten symbols 0, 1, 2, $\cdots$, 8, 9 has a unique significance/intrinsic value

2. The place at which it is located has also a significance/place value

From the available historical evidence it seems that this Hindu positional system was certainly completed at the time of eminent astronomer Brahmagupta (around 650 A.D.).

Although this achievement of human race has been narrated rather briefly in a simple manner, the course of actual historical development had many twists and turns. In a large span of time many experiments have been performed in a not so organised manner through which different number systems were born. These methods kept on competing among themselves in a subtle way. Some of them achieved temporary success but faded away with the emergence of new methods. In this way, the best number system — the decimal system could achieve a permanent place in mathematics. The actual historical development of numbers is not our concern. *However, we should not at the same time project mathematics as something which was born with perfection and was replete with intricate definitions and intriguing techniques from the very beginning. If we do so it will become detached from the real world and also become hard to appreciate. Our aim is to trace the main course of development of the number system without getting lost in the whirlpool of historical events.*

Mathematics did not rest after achieving success in symbolising the quantitative world in the form of numbers. In the daily life of primitive herdsmen, four wild sheep would join with the flock of eight domesticated sheep; sometimes five baby sheep would be born. Sometimes these herdsmen would lose six of their twenty hunting dogs while chasing some wild animals. The experience of symbolising the quantitative world inspired men to keep record of these increases and reductions in the symbolic world of numbers. So came addition, came subtraction.

1.2   Extension of Number Field

In course of time, after addition and subtraction, other mathematical operations, viz. multiplication, division, involution (raising power), evolution (lowering power) and logarithmation (taking logarithm of a number) were introduced. For quite a long time human race used natural numbers 0, 1, 2, $\cdots$ which they obtained by directly symbolising the objects of the physical world. We may represent them geometrically as points on a straight line as shown in Figure 1.1.

We may start measuring distances from a starting point O. To

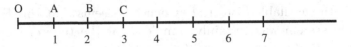

Figure 1.1: The number line

reach point A we may need to take one step, to reach B two steps and so on. We may now categorise seven algebraic operations into two broad groups — Original operations and Inverse operations as shown in Table 1.4.

In case of addition, multiplication and involution, when the operands are integers, it is guaranteed that the results will also be integers. For inverse operations it is, however, not always true. By number, if we only mean positive integers, we fail to get answers for the following cases. (a) $2 - 5 = ?$ (b) $2 \div 5 = ?$ (c) $5^{\frac{1}{2}} = ?$ and (d) $\log_2 5 = ?$ By introducing negative numbers we make (ii) possible for all integer type of operands. By introducing fractions (rational numbers) as new numbers we make (iv) viable for all integer type of operands. By introducing the concept of irrational numbers we make (vi) and (vii) valid for all integer type of operands.

1.2.1 Negative Number

We know that we can take two cows from a group of five cows. So we consider $5 - 2 = 3$ as a valid mathematical statement. We can not take five cows from a group of two cows. So $(2 - 5)$ seems to be an absurd mathematical operation. However, we may think geometrically [refer Figure 1.2] and try to find out a solution for this apparently absurd mathematical operation.

From starting point O if we move 2 steps in the forward direction

Table 1.4: Mathematical Operations

Original Operations	Inverse Operations
(i) Addition $5 + 2 = 7$	(ii) Subtraction $2 - 5 = ?$
(iii) Multiplication $5 \times 2 = 10$	(iv) Division $2 \div 5 = ?$
(v) Involution $5^2 = 25$	(vi) Evolution $5^{\frac{1}{2}} = ?$
	(vii) Logarithm $\log_2 5 = ?$

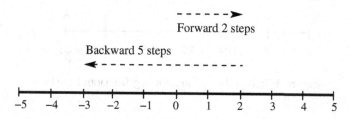

Figure 1.2: Bi-directional Number Line

we will reach B. Then we may think of moving five steps backward. If there is no wall at the starting point O this movement will be possible and finally we will reach at a point A which is 3 steps away from O in the backward direction. The distances on the left side of O are marked with -ve (negative) sign. So by extending the number line of Figure 1.1 in the backward direction and by marking the distances as $-1, -2, \cdots$ as shown in Figure 1.2 we validate subtraction of a larger number from a smaller number.

We may cite another example in this regard. If the present temperature is 2°C above freezing point and the temperature falls by 5°C it will be 3°C below freezing point. Where there are two opposite directions — forward and backward, rise and fall, left and right we may meaningfully apply negative numbers. On the contrary, when our activities are not related to directions (e.g. counting objects) negative numbers become meaningless. At this point it would be worthwhile to quote Steinmetz [2].

"In any case, however, we must realise that the negative number is not a physical, but a mathematical conception, which may find a physical representation, or may not, depending on the physical conditions to which it is applied. $\cdots$ We have become familiar with negative number at an earlier age, when we were less critical, and thus have taken it for granted, and become familiar with it by use, and usually do not realise that it is a mathematical conception, not a physical reality. When we first learned it, however, it was quite a step to become accustomed to saying, $5 - 7 = -2$, and not simply $5 - 7$ is impossible."

Indian mathematicians [notable among them are — Aryabhatta (born 476 A.D.) and Brahmagupta (born 598 A.D.)] recognised 'negative numbers' as new numbers at an early stage and denoted them

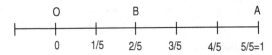

Figure 1.3: Number Line showing fractional parts

by putting a dot above the number. They named negative number with a word associated with 'debt' thereby attaching it with real life activities. The attitude of sixteenth and seventeenth century European mathematicians in this regard has been aptly described in [3] as follows.

"Negative numbers were looked upon with suspicion and expressions like 'false', 'fictitious', 'impossible' were applied to them. Such an excellent algebraist as Francois Viete (1540 -1605) rejected negative roots of equation. The great mathematician and philosopher Descartes (1596 -1650) had little understanding of the nature of negative numbers and tried to avoid them. This changed a generation later with Wallis (1618-1703) and Newton (1642 -1727) $\cdots$ many mistaken notions concerning the reality and non-reality of numbers have survived up to our days, due to the misunderstanding that numbers are identified with physical objects or properties, instead of considering them as mere symbols which can be associated with an infinite variety of physical situations."

1.2.2   Fraction and Rational Number

To carry out division operation for all numbers we need to recognise fraction as number along with integers. We may divide ten horses among 5 people. $10 \div 5 = 2$. So each of them will get 2 horses. We cannot divide 1 horse among 5 people, because in that case none of them will get any horse but each of them can get some pieces of carcass. But let us take another example similar to this one so that we can think geometrically.

Suppose we need to divide 1 metre rope among 5 people. So we may lay it on the ground and we may mark one end as O and other end as B (Refer Figure 1.3). We can divide the rope into 5 equal parts and we may mark the rope 1/5,2/5,3/5,4/5,5/5 (or 1). We may even divide the rope unequally between two people. We may give OB part (=2/5) of the rope to one person and BA part (=3/5) of the rope to another. Fractions can also be considered as the ratio of two integers

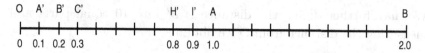

Figure 1.4: Number Line showing decimal fractional parts

e.g., 2 and 5 in case of 2/5. As such, these are called rational numbers. It is here worthwhile to indicate that all integers can also be expressed as the ratio of that integer and 1. So 4 can be expressed as 4/1. So we can introduce a type of number called rational number which includes fractions and integers. It is also important to note that a fraction, like a negative number, is basically a mathematical conception. It is applicable to those physical quantities which can be divided into smaller units. It is not applicable to those which are indivisible or lose any utility when divided into fragments. Rational numbers are also called commensurable (measurable by the same standard) numbers since both the numerator and denominator are expressed with respect to same unit. Later on we will see that there are numbers which can not be expressed as the ratio of two integers and they are not commensurable numbers. We call them irrational or incommensurable numbers.

1.2.2.1   Decimal Fraction

The decimal system was first introduced for integers only. For integers we start with 1 and count up to 9. The group 10 then follows; starting from group 1 we count up to group 9. The group-100 then follows. We may continue it indefinitely. In a similar way we can represent a quantity smaller than one.

We may again start thinking geometrically. In Figure 1.4, O is the starting point. With 1 step we reach at A and the distance OA is marked as 1. With another step we reach at B where the distance OB is 2. We may, however, divide the distance OA into 10 equal parts yielding the numbers 0.1 (OA'), 0.2 (OB'), 0.3 (OC'), $\cdots$ 0.8(OH'), 0.9 (OI'). We may further divide the distance OA' into 10 equal parts to obtain

$$0.01, 0.02, 0.03, 0.04, \cdots 0.09$$

We may further divide the distance $A'B'$ into 10 equal parts and measuring the distance from O we obtain

$$0.11, 0.12, 0.13, 0.14, \cdots 0.19$$

In this manner we can keep dividing segments $B'C'$, $C'D'$, $\cdots$, $I'A$ to obtain

$$0.21, 0.22, 0.23, 0.24, \cdots 0.29$$
$$0.31, 0.32, 0.33, 0.34, \cdots 0.39$$
$$\cdot$$
$$\cdot$$
$$\cdot$$
$$0.91, 0.92, 0.93, 0.94, \cdots 0.99$$

We can always keep on continuing this process indefinitely. No matter how close the end point is from the starting point we can always divide the distance into ten equal parts in our imagination. Using this never ending process we can generate smaller and smaller number. Earlier using the concept of grouping of objects we could generate larger and larger number without an end. Now using the concept of division we can construct smaller and smaller numbers without an end.

Now let us consider a number which has both integral and fractional parts

Ten	unit	1/10	1/100	1/1000
3	7.	9	2	8

The above symbols tell us that we have 3 packets (each packet contains ten objects), 7 loose objects, 9 broken objects each having a size of one-tenth of an object, 2 broken objects each having a size of one-hundredth of an object and 8 broken objects each having a size of one-thousandth of an object.

1.2.2.2 Conversion of fractions into decimal fractions

Only a limited group of decimal fractions having in their denominator the number 2, its multiples, and/or the number 5, its multiples can be converted into decimal fractions.

Example:

$$\frac{3}{4} = \frac{3 \times 25}{4 \times 25} = \frac{75}{100} = 0.75$$

$$\frac{2}{5} = \frac{2 \times 2}{5 \times 2} = \frac{4}{10} = 0.4$$

Since we can subdivide 1 into 10, 100, 1000 parts $\cdots$ we obtain $\frac{1}{10}(= 0.1)$, $\frac{1}{100}(= 0.01)$, $\frac{1}{1000}(= 0.001) \cdots$ However, simple fractions, like $\frac{1}{3}$ or $\frac{1}{7}$, obtained by subdividing the line segment OA (refer Figure 1.4) representing unity into 3 equal parts and 7 equal parts respectively will never coincide with any decimal subdivision however small it may be.

$$\frac{1}{3} = 0.3333 \cdots$$
$$\frac{1}{7} = 0.14257142571 \cdots = 0.\overline{14257} \cdots$$

Here we note that either a single digit (e.g., 3 in case of $\frac{1}{3}$) or a group of digits (e.g., 14257 in case of $\frac{1}{7}$) keeps on repeating in the decimal part without any end. So we observe that although a decimal fraction can represent smaller and smaller numbers without any end, it fails to represent rather simple fractions using finite number of digits. The world of small numbers is really very intriguing. Afterwards we will see that there are yet other numbers which can not even be represented as the ratio of two integers. An attempt to obtain decimal representation of such numbers yields non-terminating, non-repeating decimal digits. These numbers are called Irrational Numbers. It is very important to note that in spite of its capacity of constructing larger and larger numbers and smaller and smaller numbers, decimal system is not omnipotent and provides us no panacea.

1.2.3 Irrational Number

Cornelius Lanczos nicely summed up [7] the effect of introduction of negative numbers and fractions in our number system.

"With the creation of negative numbers and fractions, the number field is closed with respect to the four fundamental operations of arithmetic: addition, subtraction, multiplication and division. None of these operations applied any number of times will ever lead us out of the field. The sum and difference, product and ratio of any two fractions can again be written as a fraction (with positive and negative sign). Hence we can not encounter a situation which would demand

a further enlargement of the number system. $\cdots$ *However, there are situations which are not covered by the four basic operations."*

Let us now consider Involution (raising power) and Evolution (lowering power). If we multiply a number several times by itself we say that involution is being carried out.

$$5 \times 5 \times 5 \times 5 = 625$$

or, $5^4 = 625$

In general, $p^m = q$, here p and m are positive integers (we can extend this idea when these are positive or negative fractions). It is also guaranteed that q will be a positive integer.

Next if we ask to find the number which when multiplied by itself four times produces 625, one needs to perform evolution. Mathematically it is expressed as

$$625^{\frac{1}{4}} = \sqrt[4]{625} = 5$$

In general, $q^{\frac{1}{m}} = \sqrt[m]{q} = r$. In case of evolution we fail to obtain any rational number for some positive integer values of r and m.

We have listed below two sequences (natural numbers and their squares):

Natural Numbers :	1	2	3	4	5
Square of natural numbers:	1	4	9	16	25

We know that $1^2 = 1$ and $2^2 = 4$. Therefore, $\sqrt{4} = 2$.

We guess that $\sqrt{2}$ will be a number which is greater than 1 but less than 2.

1.2.3.1   Proof of irrationality of $\sqrt{2}$

Let us assume that,

$$\sqrt{2} = \frac{a}{b} \tag{1.1}$$

where a and b are integers and they have no common factor.

Let us now note the following statements.

1. Square of an even integer is even.

2. Square of an odd integer is odd, because

$$(2n+1)^2 = 4n^2 + 4n + 1 = 4n(n+1) + 1$$

Squaring equation (1.1) we get $2 = \frac{a^2}{b^2}$

$$\therefore \quad a^2 = 2b^2$$

Since a^2 is even a is also even. Let $a = 2c$

$$\therefore \quad 4c^2 = 2b^2$$
$$\therefore \quad b^2 = 2c^2$$

As such, b is also even.

Since a and b are both even they have a common factor 2. However, this contradicts our initial presumption that a and b do not have any common factor. As such, $\sqrt{2} \neq \frac{a}{b}$. In other words $\sqrt{2}$ can not be a rational number. So we introduce a term *Irrational Number* and call $\sqrt{2}$ an irrational number.

1.2.3.2 Proof of irrationality of $\sqrt{3}$

First of all we should note that the square of an integer is divisible by 3 if and only if the integer itself is divisible by 3. Let n be an integer; so the number $3n$ is divisible by 3.

$$(3n)^2 = 9n^2 = 3.(3n^2),$$

hence the square of $3n$ is divisible by 3.

The number $3n + 1$ is not divisible by 3.

Now,

$$(3n + 1)^2 = 9n^2 + 6n + 1 = 3n(3n + 2) + 1,$$

It is not divisible by 3.

The number $3n + 2$ is not divisible by 3.

Now,

$$(3n + 2)^2 = 9n^2 + 12n + 4 = 3(3n^2 + 4n + 1) + 1.$$

It is not divisible by 3.

Let us assume that,

$$\sqrt{3} = \frac{a}{b} \tag{1.2}$$

where a and b are integers and they have no common factor.

Squaring equation (1.2) we get $3 = \frac{a^2}{b^2}$

$$\therefore \quad a^2 = 3b^2$$

Since a^2 is divisible by 3, a is also divisible by 3.

Now, let $a = 3c$

$$\therefore \quad 9c^2 = 3b^2$$

$$\therefore \quad b^2 = 3c^2$$

$$\therefore \quad \text{b is also divisible by 3}$$

Since both a and b are divisible by 3 they have a common factor 3. However, this contradicts our initial presumption that a and b do not have any common factor. As such, $\sqrt{3} \neq \frac{a}{b}$.

If we make an attempt to represent $\sqrt{3}$ and $\sqrt{2}$ in decimal form, we get $1.73205080\cdots$ and $1.41421356\cdots$ respectively. So in both cases we get *non-terminating, non-repeating decimal fractional part*. In general, a number with a non-terminating, non-repeating decimal fractional part represents an irrational number.

It is also interesting to note that it is possible to represent irrational numbers as a continued fraction.

1.2.3.3   Representing irrational numbers as continued fractions

Method 1

Since $\sqrt{2}$ lies between 1 and 2, we may write

$$\sqrt{2} = 1 + \frac{1}{y} \tag{1.3}$$

where y is greater than 1.

$$\therefore \quad \frac{1}{y} = \sqrt{2} - 1$$

$$\therefore \quad y = \frac{1}{\sqrt{2}-1} = \frac{\sqrt{2}+1}{2-1} = 1 + \sqrt{2} \tag{1.4}$$

Adding 1 to both sides of equation (1.3) we get

$$1 + \sqrt{2} = 2 + \frac{1}{y}$$

Using equation (1.4) we can write

$$y = 2 + \frac{1}{y} \tag{1.5}$$

Using equation 1.3 and equation (1.5) we can write

$$\sqrt{2} = 1 + \cfrac{1}{2 + \frac{1}{y}} \tag{1.6}$$

We may also write

$$\sqrt{2} = 1 + \cfrac{1}{2 + \cfrac{1}{2 + \frac{1}{y}}} \tag{1.7}$$

We can go on and on repeating substitution of the value of y this way.

In equation (1.6), if we neglect $\frac{1}{y}$, we get $\sqrt{2} \approx \frac{3}{2}$

In equation (1.7), if we neglect $\frac{1}{y}$, we get $\sqrt{2} \approx \frac{7}{5}$

$$\therefore \quad \frac{7}{5} < \sqrt{2} < \frac{3}{2} \qquad \text{Range} = \frac{3}{2} \cdots \frac{7}{5} = 0.1$$

In this way, on further substitution of the value of y we may express $\sqrt{2}$ approximately by rational numbers with ever increasing accuracy and narrow the range in which it lies.

$$\frac{41}{29} < \sqrt{2} < \frac{17}{12} \qquad \text{Range} \approx 0.0028736$$

$$\frac{239}{169} < \sqrt{2} < \frac{99}{70} \qquad \text{Range} \approx 0.0000845$$

Method 2

$$(\sqrt{y} - 1)(\sqrt{y} + 1) = y - 1$$

$$\therefore \quad \sqrt{y} = 1 + \frac{y-1}{1+\sqrt{y}}$$

Let us now put $y = 2$

$$\therefore \quad \sqrt{2} = 1 + \frac{2-1}{1+\sqrt{2}} = 1 + \cfrac{1}{1+1+\frac{1}{1+\sqrt{2}}} = \cdots$$

We can go on and on repeating substitution of the value of $\sqrt{2}$ this way.

Method 3

$$\text{Let } (1 + r)^2 = 2$$

$$\therefore \quad 1 + r^2 + 2r = 2$$

$$\therefore \quad r(r + 2) = 2 - 1 = 1$$

$$\therefore \quad r = \frac{1}{2+r} = \frac{1}{2+\frac{1}{2+r}} = \cdots$$

$$\sqrt{2} = 1 + r = 1 + \frac{1}{2+r} = 1 + \frac{1}{2+\frac{1}{2+r}} = \cdots$$

1.2.3.4   Transcendental Numbers

It is, however, important to note that although we can not express irrational numbers like $\sqrt{2}$ and $\sqrt{3}$ as a ratio of two integers they satisfy algebraic equations $x^2 - 2 = 0$ and $x^2 - 3 = 0$ respectively. Obtaining the values of the numbers is not the issue here; the issue is how to define them in terms of integers. It is worthwhile to note that the equations mentioned here have integral coefficients. Rational number like $\frac{2}{11}$ satisfy linear equation $11x - 2 = 0$. In a similar fashion $\sqrt{2}$ satisfies the quadratic equation $x^2 - 2 = 0$. In this way we can define a large group of irrational numbers by solving even higher degree polynomials with integral coefficients. This type of irrational numbers is termed as *algebraic* type of irrational numbers. However, there are numbers which can not be defined this way and they do not satisfy any polynomial with integral coefficients. So it seems that they transcend the power of algebra. As such, they are called *transcendental numbers*. π, e, $3^{\sqrt{2}}$, $\sqrt{3}^{\sqrt{2}}$ are examples of transcendental numbers. Soviet mathematician Gelfand proved that all numbers of the form α^β are transcendental numbers provided α is neither zero nor 1 and β is an irrational number.

1.2.4   Imaginary (or Quadrature Number)

$$\text{We know that } (+2) \times (+2) = 4$$

$$\text{Again} (-2) \times (-2) = 4$$

So, $\sqrt{4}$ has two values $+2$ and -2. Square of a number (we have encountered so far) is always positive irrespective of the sign of the number. As such, it appears that it is futile to search for a value of $\sqrt{-4}$. However, we may again think geometrically to find a solution to this problem.

Let us take a line section OA (Figure1.5) which represents the number 2. If we multiply 2 by -1 we get (-2) represented by line section OB. We may now infer that *multiplication by -1 causes a rotation of $180°$ in the anti-clockwise direction*. If we further multiply

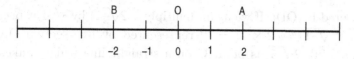

Figure 1.5: Number line showing positive and negative integers

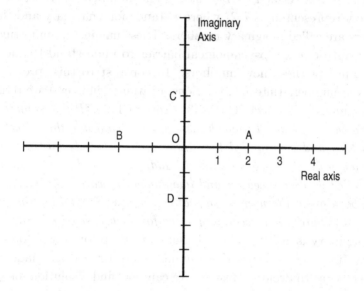

Figure 1.6: Real axis and imaginary axis in Number Plane

(−2) by −1, we get (+2) represented by line section OA. So now the line section OB undergoes a further rotation of 180° in the anti-clockwise direction and becomes OA. Now for the time being let us postpone our search for the significance of the square root of a negative number and merely accept $\sqrt{-1}$ as a symbol which when multiplied by itself produces (−1). Now if we multiply 2 by by $\sqrt{-1}$ we get $2\sqrt{-1}$ [$= \sqrt{-2}$]. If we further multiply by $\sqrt{-1}$ we get $2 \times \sqrt{-1} \times \sqrt{-1} = 2 \times -1 = -2$. So we observe that if we multiply by $\sqrt{-1}$ *twice* it causes a rotation of 180° in the anti-clockwise direction. We may also infer that if we multiply by $\sqrt{-1}$ only once it will cause a rotation of 180° ÷ 2 or 90° in the anti-clockwise direction. We may now draw a perpendicular at the point O (refer Figure 1.6). The line OC lying on this perpendicular may be marked as $2\sqrt{-1}$ and OD may be marked as $-2\sqrt{-1}$.

If we multiply (−2) [represented by OB] by $\sqrt{-1}$ it becomes (−2$\sqrt{-1}$)

[represented by OD]. If we again multiply $(-2\sqrt{-1})$ by $\sqrt{-1}$ it becomes $[-2\sqrt{-1} \times \sqrt{-1} = -2 \times -1 = 2]$ represented by line section OA. So we may think $2\sqrt{-1}$ as point C on a straight line which makes $90°$ with the straight line on which line section OA lies. The straight line on which is termed as *Imaginary axis* and the straight line on which A lies is termed as *real axis*. The numbers $\sqrt{-1}$, $2\sqrt{-1}$ are concisely represented as $i, 2i$ where i stands for imaginary and these numbers are called imaginary numbers. These naming are unfortunate as they introduces a psychological barrier to understand the actual significance of these new numbers. In contrast to this, previously conceived numbers (integers, fractions, irrational numbers) are termed as real numbers. As such, Gauss [3] wrote in 1831, *"If this subject has hitherto been considered from the wrong view point and thus enveloped in mystery and surrounded by darkness, it is largely an unsuitable terminology which should be blamed. Had $+1, -1$, and $\sqrt{-1}$, instead of being called positive, negative and imaginary (or worst still impossible) unity, been given the names, say, of direct, inverse and lateral unity, there would hardly have been any scope for such obscurity."* Since the tag imaginary is still being used, the misconception lingers on even to-day. In some literature these numbers are termed as quadrature numbers so as to avoid confusion. We can now find a solution for $\sqrt{-4}$.

$$\sqrt{-4} = \sqrt{4 \times -1} = \sqrt{4}.\sqrt{-1} = 2i$$

1.2.4.1   Complex (Directed or General) number

We assume that all positive and negative numbers are in one-to-one correspondence with the points on the real axis shown as a horizontal line in Figure 1.7 and could be considered as distances from the point **O**. Likewise we assume that all positive and negative imaginary numbers are in one-to-one correspondence with the points on the imaginary axis drawn vertically in Figure 1.7 and could be considered as distances from the point **O** in vertically upward and downward directions.

We may now consider other points (on the same plane in which the real axis and imaginary axis belong) which lie neither on the real axis nor on the imaginary axis. We may reach such a point from the point **O** by first moving certain distance along the real axis and then moving certain distance along a line parallel to the imaginary axis.Thus we may reach point P1 by moving a distance

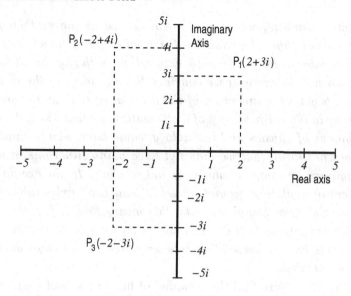

Figure 1.7: Pictorial Representation of Complex Numbers

$OA(2)$ along real axis and then a distance $AP_1(= OQ_1 = 3i)$ along a line which is parallel to the imaginary axis. We associate the point P_1 with a new type of number which is expressed by combining the two distances using a $+$ sign i.e. as $(2 + 3i)$. Likewise P_2 and P_3 can be associated with the complex numbers $(-2 + 4i)$ and $(-2 - 3i)$ respectively. This type of representation of a complex number is known as rectangular or Cartesian representation. We may represent it also in polar form where we directly find the distance of the line section $OP_1(= \sqrt{((OA)^2 + (OQ_1)^2)})$ and call it the modulus of the complex number. We the find the angle θ_1 the line section OP_1 makes with the positive direction of real axis $(\theta_1 = \tan^{-1}\frac{OQ_1}{OA})$ and call it the argument of the complex number. So the point P_1 corresponds to a complex number $3.6\angle 56.31°$. At this point it would be worthwhile to see how mathematician C. Lanczos [4] assessed the impact of invention of so called 'imaginary' and 'complex' numbers in mathematics.

"Such 'vectors' are eminently important from the standpoint of physics where directed quantities appear with great regularity. Velocity, acceleration, force, and other physical entities are all representable by vectors. It is true that these vectors usually sweep over a three-dimensional space rather than a plane. Yet in the many problems of physics which are restricted to a plane, we have immediate use

of complex numbers, which can be considered as the mathematical counterparts of physical quantities. Far from anything 'imaginary', the complex numbers have innumerable associations with physical situations and we do not go wrong if we consider the discovery of the complex numbers, beyond the invention of the decimal system, as the greatest eye-opener in the entire history of mathematics, without which the rapid development of physics and technology would have been unthinkable. This does not mean that we may not have many situations in which an imaginary or complex number is out of place. If, for example, in a problem of analytical geometry we find that two circles intersect in some imaginary or complex points, this means that in fact these two circles do not intersect. It depends always on the special circumstances which decide how an association between numbers and physical reality has to be established."

Let us now try to find the meaning of higher roots of (-1). As for example we may consider $\sqrt[4]{-1}$; it is such a number which produces -1 (i.e. causes a rotation of $180°$) when we multiply 1 by this number 4 times. So when we multiply 1 by this number only once it should cause a rotation of $180° \div 4 = 45°$ (in polar form).In Cartesian form we express it as $(\frac{1}{\sqrt{2}} + \frac{1}{\sqrt{2}}i)$. In Fig. 1.8 it is located at M_1. However, we can find three more such numbers viz. $1\angle135°$ (Since $4 \times 135° - 360° = 540° - 360° = 180°$), $1\angle225°$ (Since $4 \times 225° - 2 \times 360° = 900° - 720° = 180°$) and $1\angle315°$ (Since $4 \times 315° - 3 \times 360° = 1260° - 1080° = 180°$). In Cartesian form we express them as $(\frac{-1}{\sqrt{2}} + \frac{1}{\sqrt{2}}i)$, $(\frac{-1}{\sqrt{2}} - \frac{1}{\sqrt{2}}i)$ and $(\frac{1}{\sqrt{2}} - \frac{1}{\sqrt{2}}i)$. In Figure 1.8 these are located at M_2, M_3 and M_4.

1.2.5 Quaternion

Hamilton introduced this number and related mathematics in 1843. Literary it means 'a set of four'. While using this name to denote a new mathematical concept he was greatly influenced by Tetractys, the ancient sacred Pythagorean symbol which consists of ten dots arranged in four rows. By the introduction of imaginary number and complex number, application of system of numbers was extended from line (a one-dimensional quantity) to plane (a two-dimensional quantity). So mathematicians put great effort to devise numbers which could be applied to three dimensional quantities related to space. Ordinary complex number has a form $a + bi$; as such, for three dimensional quantities Hamilton suggested a form $a + bi + cj$ where $i = j = \sqrt{-1}$.

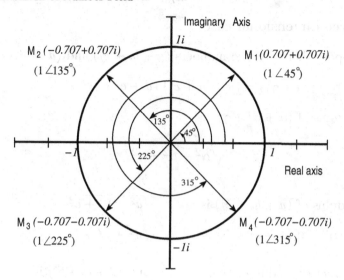

Figure 1.8: Roots of $\sqrt[4]{-1}$ in the complex plane

This form suits well for addition. However, he failed to multiply two such numbers fulfilling the *law of qqmoduli* [3] even after struggling with the problem for quite a long time. The *law of moduli* states that the product of moduli of two numbers is equal to the modulus of the product of two numbers.

Illustration

Two dimensional case

Suppose there are two numbers $(a + bi)$ and $(c + di)$.

Modulus of $(a + bi)$ is $\sqrt{a^2 + b^2}$

Modulus of $(c + di)$ is $\sqrt{c^2 + d^2}$

Now, $(a + bi).(c + di) = (ac - bd) + (bc + ad)$

Modulus of the product is $\sqrt{(ac - bd)^2 + (bc + ad)^2}$

$$= \sqrt{a^2c^2 + b^2d^2 - 2abcd + b^2c^2 + a^2d^2 + 2abcd}$$

$$= \sqrt{a^2(c^2 + d^2) + b^2(c^2 + d^2)}$$

$$= \sqrt{(a^2 + b^2)}.\sqrt{(c^2 + d^2)}$$

So we find that law of moduli holds good for numbers which represent two dimensional quantities.

Three dimensional case

Suppose there are two numbers $(a + bi + cj)$ and $(d + ei + fj)$

Let $(\alpha + \beta i + \gamma j) = (a + bi + cj).(d + ei + fj)$

Modulus of $(\alpha + \beta i + \gamma j)$ is

$$\sqrt{\alpha^2 + \beta^2 + \gamma^2} \qquad (1.8)$$

Modulus of $(a + bi + cj)$ is $\qquad \sqrt{a^2 + b^2 + c^2}$

Modulus of $(d + ei + fj)$ is $\qquad \sqrt{d^2 + e^2 + f^2}$

Product of moduli of $(a + bi + cj)$ and $(d + ei + fj)$ is

$$\sqrt{a^2 + b^2 + c^2}.\sqrt{d^2 + e^2 + f^2} \qquad (1.9)$$

Hamilton failed to develop an algebra to make equation (1.8) and equation (1.9) equal.

Afterwards he realised that to represent a point in space one needs to set up three mutually perpendicular axes viz. x, y and z and treat them in equal manner. So, three imaginary units viz. i, j and k rather than two viz. i and j need to be introduced. As such, he suggested a number of the form $(a + bi + cj + dk)$ and assume

$$i^2 = j^2 = k^2 = ijk = -1$$

$$ij = k; \quad jk = i; \quad ki = j$$

$$ji = -k; \quad kj = -i; \quad ik = -j$$

It is possible to make *law of moduli* applicable. *It is, however, important to note that since $i.j \neq j.i$ this new algebra does not obey commutative law of algebra for multiplication.* It was a bold step in a sense that he had the courage to go beyond the tradition while introducing Quaternion.

Hamilton thought that quaternion is in harmony with our physical world of space and time. The real number a may mean time and b, c and d may mean length, breadth and height respectively. He once wrote "Time is said to have only one dimension, and space to have three dimensions $\cdots$ The mathematical quaternion partakes of both these elements; in technical language it may be said to be 'time plus space', or, 'space plus time': and in this sense it has, or at least involves a reference to, four dimensions. And how the One of Time, of Space the Three, Might in the Chain of Symbols girdled be." (Graves, Robert Percival (1889) Life of Sir William Rowan Hamilton, Knt., ll.d., d.c.l., m.r.i.a., Andrews professor of astronomy in the university of Dublin, and Royal astronomer of Ireland, etc.: Including selections from his poems, correspondence, and miscellaneous writings.Vol. III., Dublin: Dublin University press series.)

In spite of its mathematical merits physicists and engineers considered quaternion as a complicated tool which is difficult to use for physical applications. As such, American physicist Josiah Willard Gibbs (1839 — 1925) proposed a modified version where he considered only the vector part $(bi + cj + dk)$ of a quaternion and replaced quaternion multiplication by two types of multiplication, scalar or dot product and vector or cross product. He termed his method "Vector Analysis". English Communication engineer Oliver Heaviside (1850–1925) also made important, independent contribution in this regard. Physicists and engineers found it to be much simpler than quaternion. So Hamilton's quaternion went into oblivion for quite sometime. Quaternion, however, made a come back in the late twentieth century. Since it can handle three dimensional rotations better than other existing methods it is now used in Computer Vision, Computer Graphics, Avionics and Robotics. In Physics it is used in the form of Pauli matrices in the introduction to quantum mechanics course.

1.2.6 Infinitesimal and Hyper-real Numbers

Infinities and indivisibles transcend our finite understanding, the former on account of their magnitude. the latter because of their smallness; Imagine what they are when combined.

— Galileo Galilei as Salviati
in Dialogues Concerning Two New Sciences (1638)

1.2.6.1 Background

The emergence of calculus is considered to be a turning point in the history of mathematics. Gottfried Wilhelm Leibniz (1646 –1716) and

Isaac Newton (1642–1727) independently "invented" calculus in the
later part of seventeenth century by unifying a collection of isolated
results and systematically organising them as a whole. However, the
logical foundation of calculus was not very sound at that time.

"*Newton, at different times, described the derivative of y (which
he called the 'fluxion' of y) in three different ways, roughly*

1. *The ratio of an infinitesimal change in y to an infinitesimal change
   in x. (The infinitesimal method)*

2. *The limit of the ratio of the change in y to an change in $x, \frac{\Delta y}{\Delta x}$, as
   Δx approaches zero. (The limit method)*

3. *The velocity of y where x denotes time. (The velocity method)*

   *In his later writings Newton sought to avoid infinitesimals and
emphasized method (2) and (3).*

   *Leibniz rather consistently favoured the infinitesimal method but
believed (correctly) that the same results could be obtained using only
real numbers. He regarded the infinitesimals as 'ideal' numbers like the
imaginary numbers.*" [7]

The weakness of the infinitesimal based method used by Leibniz
can be demonstrated with the help of the following example.

Let us consider a function $y = f(x) = x^2$

To find the derivative of y with respect to x, we consider a small
change in x denoted by the symbol Δx and calculate the following

$$
\begin{aligned}
\frac{f(x + \Delta x) - f(x)}{\Delta x} &= \frac{(x + \Delta x)^2 - x^2}{\Delta x} \\
&= \frac{x^2 + (\Delta x)^2 + 2x\Delta x - x^2}{\Delta x} \\
&= 2x + \Delta x
\end{aligned}
\tag{1.10}
$$

Considering Δx to be an infinitesimal it is now neglected. The
ratio defined by the L.H.S. above is termed as the derivative of y with
respect to x and it is considered to be equal to $2x$. However, the
removal of Δx in the final step is done rather in an arbitrary manner
without following any definite principle. Removal of Δx is justified
only if it is equal to zero. In that case, however, we can not execute
division by Δx. Bishop Berkeley in his famous book *The Analyst, Or*

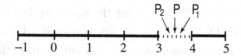

Figure 1.9: Hyper-real line

a *Discourse Addressed to an Infidel Mathematician*, made a scathing attack against such practices which fail to provide any consistent logic in support of this method.

In 1870's Karl Weierstrass (1815-1897) established Calculus on the sound footing of Limits which was defined in terms of '$\epsilon - \delta$' condition. He also constructed 'real number system' from the positive integers during this time. The method followed by Weierstrass is known as 'Standard Method'. Weierstrass suggested that one should take the limit of the above mentioned ratio as Δx tends to zero.

Much of the Calculus developed from its birth till 1870s by stalwarts like the Bernoullis, Euler, Lagrange and Cauchy (to name a few), however, were based on the intuitive concept of infinitesimals. From the time the concept of limit was introduced, infinitesimals have been banished from the formal Calculus course for reason of mathematical rigor. Students, now, have to learn the subject without the original intuition. The notion of infinitesimals is deeply ingrained in our mind; we frequently use words like 'instantaneous velocity' since we know when a moving car hits a wall the severity of the crash depends on the velocity at the instant it hits the wall. Abraham Robinson in 1960s found a way to make infinitesimal rigorous by using a method known as Non-standard Analysis in which a new type of number called Hyper Real number was introduced. Therefore, hyper real numbers enable us to make a marriage between intuition and rigor while we deal with Calculus.

1.2.6.2 Basic Ideas

We have noticed earlier that we can set a one-to-one correspondence between points on a straight line and real numbers. Suppose a particular point 'P' represents the real number 3.14 (see Figure 1.9).

Now, if any one asks us to show the point 'P_1' which is on the right side of point 'P' and closest or nearest to it is no definite answer to it. In a similar way if somebody wants to know about the nearest

left side neighbour (say 'P_2') of point 'P' again we fail to give him/her
a definite answer. The microscopic world around 'P', representing a
real number, is a strange one. We may look at it through a microscope
knowing quite well that its power of magnification can be increased
indefinitely. Every time we increase the power of magnification of
the microscope we find new 'P_1' and 'P_2'. So we take resort to
imagination while remaining alert that our imagination does not lead
us to any contradiction or absurdity. This demands a sound knowledge
in mathematical logic. In spite of the fact that the author of this book
does not have any expertise in this regard, effort will be made to make
the readers acquainted with the basic ideas in the following paragraph.

In Standard Analysis it is assumed that all neighbours of point 'P'
represent real numbers. In Non-Standard Analysis it is imagined that
there are two groups of neighbouring points around 'P' representing
viz. hyper real numbers other than real numbers and real numbers;
neighbouring points representing hyper real numbers which are not
real numbers lie nearer to 'P' when compared with the neighbouring
points which represent real numbers. These hyper real numbers may
be represented by $(3.14 + \epsilon)$ and $(3.14 - \epsilon)$ where ϵ is an infinitesimal.
An infinitesimal ϵ is infinitely close to zero although it is not equal
to zero. ϵ is smaller than any +ve real number but greater than any
−ve real number. $\frac{1}{\epsilon}$ gives us a positive infinite number and $\frac{1}{-\epsilon}$ gives
us a negative infinite number. In Standard Analysis a straight line
whose points represent numbers is called real axis. In Non-Standard
Analysis a straight line whose points represent numbers is called hyper
real axis. It contains points which represent real numbers (may be
termed as 'r') and other numbers like '$r + \epsilon$', '$r - \epsilon$', ' ϵ', ' $-\epsilon$', '$\frac{1}{\epsilon}$' and
'$\frac{1}{-\epsilon}$'. As such, real numbers become a sub set of hyper real numbers.
It is important to note that like real numbers hyper real numbers also
contain positive and negative infinitesimals and positive and negative
infinite numbers. One may, however, question the authenticity of this
statement. It would be worthwhile to quote Keisler [7] in this regard.

*"We have no way of knowing what a line in physical space is really
like. It might be like the hyper real line, the real line, or neither.
However, in application of the calculus it is helpful to imagine a line
in physical space as a hyper real line. The hyper real line is, like the
real line, a useful mathematical model for a line in physical space."*

Hyper real numbers are yet another extension of number system.
They have been extended from real numbers in order to bring in-

finitesimals and infinite numbers in the number system as legitimate members. In this regard following three principles play important role.

- The Extension Principle
- The Transfer Principle
- The Standard Part Principle

The Extension Principle provides hyper real numbers and extends all real functions to these numbers. The Transfer Principle enables us to compute with hyper real numbers in a way which is exactly same as the manner we do computation with real numbers.

Real numbers may also be termed as Standard numbers. The Standard Part Principle states that every finite hyper real number is infinitely close to only one unique real number. If a is a hyper real number then in general $a = st(a) + \epsilon$ where $st(a)$ is the real number to which a is infinitely close and ϵ is an infinitesimal number. So the hyper real number a has standard (real) part and an infinitesimal part. If a is a real number then the infinitesimal part will be absent and $a = st(a)$. Infinite hyper real numbers do not have standard parts.

Considering Δx in equation (1.10) to be an infinitesimal Robinson defined derivative of y with respect to x as the Standard Part of $\frac{f(x+\Delta x)-f(x)}{\Delta x}$. As such

$$st\left(\frac{f(x+\Delta x)-f(x)}{\Delta x}\right) = st\left(2x + \Delta x\right) = 2x.$$

Since x is a real number $st(2x) = 2x$ and Δx is an infinitesimal $st(\Delta x) = 0$. Let me now quote Ian Stewart "*In stead of the extra Δx being swept under the carpet with much special pleading, it is neatly expunged*".

The positive points of use of non-standard analysis based infinitesimals in Calculus are as follows.

- Infinitesimals have some intuitive appeal. So, we can combine rigor and intuition while studying calculus.
- It simplifies and shortens proofs.
- Although it was banished from calculus it has been used by physicists and engineers. It finds use in quantum mechanics, special theory and general theory of relativity etc.
- Infinitesimals have also been used in Economics.

Bibliography

1. Boyer, C. B., 1968. *A History of Mathematics.* Wiley International Edition.

2. Burton D. M., 2011. *The History of Mathematics — An Introduction.* McGraw Hill Seventh Edition

3. Kaplan R. and Kaplan E., 2003. *The Art of the Infinite — The Pleasure of Mathematics.* Oxford University Press.

4. Keisler H. J., 2000. *Elementary Calculus — An Infinitesimal Approach* Second Edition

5. Stillwell J., 2010. *Mathematics and its History.* Third Edition, Springer

6. Courant R. and Robbins H., 1996 (and revised by Stewert I.) *What is Mathematics.* Oxford University

7. Lanczos C., 1968. *Numbers Without End.* Edinburgh : Oliver and Boyd.

8. Steinmetz C. P., 1911. *Engineering Mathematics.* New York, McGraw Hill Book Company.

Chapter 2

On e & e^x

By trying and erring, by groping and stumbling — so progressed our knowledge. Hampered and yet spurred by a hard struggle for existence, a plaything of his environment and slave to the traditions of his time man was guided in this progress not by logic but by intuition and stored up experience of his race. This applies to all things human, and I have made painstaking efforts to show that mathematics is no exception.

— Tobias Dantzig

2.1 Introduction

The number e or the function e^x was not a sudden creation made by a genius. The human race took a long time to detect the entity of e in a rather veiled form in some commercial activities. Later on, people observed the presence of the same entity in different natural processes. To start with, people were not very sure whether they should call it a number. Afterwards they revised the concept of numbers to give this entity the status of a number. When a child learns to talk he/she does not face much difficulty in using words like 'fish' or 'doll'. However, it takes a longer time to grasp the actual meanings of words like 'misery' or 'mercy' or 'government'. By keenly observing a number of incidents a child gradually picks up these words. Likewise in mathematics it is easier to understand the meanings of the numbers 5 or 358 or 13.75. To appreciate the meaning of e or e^x, however, one needs to critically observe a few changing processes — both in the natural and in the social world. Gradually people came to know how the function e^x

is related to other functions like logarithmic functions, trigonometric functions (sin, cos etc.) and hyperbolic functions (sinh,cosh etc.). The function e^{mx} has a very interesting property — its derivative is proportional to the function itself. This property helps us in a significant way while we intend to solve differential equations. When the exponent x in e^x is an imaginary number we get a planar vector (whose existence is confined to the complex plane) and when this imaginary x becomes a time varying quantity ($x = \omega.t$ where t is the time and ω is a constant) we get a rotating vector. When the exponent x in e^x is a complex number linearly varying with time we get a rotating vector with varying length. All these features are widely used in many branches of science and engineering. In our pursuit of knowing the value of ∞^0 using the function $x^{\frac{1}{x}}$ by increasing the value of x from 1 to ∞ we find e provides us the point at which the function achieves maximum value. Likewise in search for the value of 0^0 using the function x^x by decreasing the value of x from 1 to 0 we find e^{-1} provides us the point at which the function achieves minimum value. The number e and the function e^x have many intriguing features; so let us delve into the realm of e and e^x.

2.2   Backdrop in which e emerged as the outcome of continuous compounding

Let me first tell you the background story of emergence of the rudimentary concept of e. During the last part of sixteenth century there was an upsurge of commercial activities in Europe. Calculation of compound interest was a practical necessity. To meet this need Stevinus, a habitant of Belgium, a financier and engineer, prepared a table for finding the compound interest and published it in 1582. Now let me explain what happens when we keep on compounding continuously. Let A has lent Rs. 1 to B on a contract that B will pay him back the principal of Rs. 1 along with 100 % annual interest at the end of one year. In such a case B will pay Rs. 2 (Rs. 1 as principal and Rs. 1 as interest) after 1 year. Let us now change the contract a little bit so that interest is calculated both at the end of 6 months and after the completion of one year. At the end of six months the interest will be 50 paise. So the principal for the next six months will be Rs. 1.50 and its interest for the last six months will be 75 paise. As such, at the end of 1 year B needs to pay an amount of Rs. 2.25 (Rs. 1.50 as principal and Rs. 0.75 as interest). So B needs to pay more this time. If we now calculate interest at the end of every 3 months the final amount that B needs to pay at the end of one year will be even more. The following table shows how the final amount increases with the increase in the number of intervals at the end of which interest is calculated repeatedly.

Table 2.1: Effect of increase in number of compounding on A_n

Number of intervals (n) in a year	Final Amount (Rs.)
1	2.000
2	2.250
3	2.370
4	2.441
5	2.488
6	2.521
7	2.546
10	2.593
100	2.704
1000	2.716
10000	2.718

One can observe from Table 2.1 as the number of intervals become larger and larger change in the final amount becomes smaller and smaller. This table can be easily developed from the well known formula used for compound interest problems $A_n = (1 + \frac{1}{n})^n$, where n is the number of intervals and A_n is the final amount.

The number e can be formally defined as

$$e = \lim_{n \to \infty} (1 + \frac{1}{n})^n$$

Although Stevinus published the Table in a book, he mentioned that French Mathematician Jean Trenchant invented these Tables earlier in 1558.

2.3   Outcome of decrease through continuous compounding

Now let us see how much will be left at the end of 1 year if an amount of 1 unit keeps on decreasing continuously at a rate of 100%. We may use the equation $A'_n = (1 - \frac{1}{n})^n$ for this purpose. Here n is the number of intervals and A'_n is the final amount.

Table 2.2: Effect of increase in number of compounding on A'_n

Number of intervals (n) in a year	Final Amount (Rs.)
1	0.000
2	0.250
3	0.296
4	0.316
5	0.328
6	0.335
7	0.340
10	0.349
100	0.366
1000	0.368
10000	0.368

As the number of interval n becomes large (refer Table 2.2) there is no appreciable change in the final amount. It is interesting to note that

$$\lim_{n\to\infty} (1 - \frac{1}{n})^n = \frac{1}{\lim_{n\to\infty}(1 + \frac{1}{n})^n} = \frac{1}{e} \approx 0.368$$

Physical example of e^{-1}:

- If an asset initially worth Rs. 1 keeps on depreciating continuously at a rate of 100% at the end of 1 year its worth reduces to e^{-1}.

- A radioactive material decays in this manner.

- "The reciprocal of the number e appears in the following problem of probability of misaddressed letters. If n letters are written to different addresses and there are n matching envelopes, then the probability of every letter being put in a wrong envelope quickly converges to the value $\frac{1}{e}(\approx 0.36787944117...$ as the number n increases. It is assumed that the letters put into envelopes randomly with uniform probability distribution." (Page no. 104, [13])

2.4 e as an infinite series

Expansion of $(1 + \frac{1}{n})^n$:

We know,

$$(1 + a)^n = 1 + na + n(n-1).\frac{a^2}{2!} + n(n-1)(n-2).\frac{a^3}{3!} + \cdots$$
$$+ n(n-1)(n-2)\ldots(n-r+1).\frac{a^r}{r!} + \cdots \infty \qquad (2.1)$$

When $a = \frac{1}{n}$,

$$(1 + \frac{1}{n})^n = 1 + n.\frac{1}{n} + n^2.\frac{(1-\frac{1}{n})}{n^2.2!} + n^3.(1-\frac{1}{n})\frac{(1-\frac{2}{n})}{n^3.3!} + \cdots \infty \quad (2.2)$$

So,

$$e = \lim_{n\to\infty}(1 + \frac{1}{n})^n = 1 + 1 + \frac{1}{2!} + \frac{1}{3!} + \cdots \infty \qquad (2.3)$$

In equation (2.2) if we put $n = -m$ we get,

$$(1 - \frac{1}{m})^{-m} = 1 + m.\frac{1}{m} + m^2.\frac{(1+\frac{1}{m})}{m^2.2!} + m^3.(1+\frac{1}{m})\frac{(1+\frac{2}{m})}{m^3.3!} + \cdots \infty$$

Now,

$$\lim_{m\to\infty}(1 - \frac{1}{m})^{-m} = 1 + 1 + \frac{1}{2!} + \frac{1}{3!} + \cdots \infty$$

$$\therefore \quad e = \lim_{m\to\infty}(1 - \frac{1}{m})^{-m} = \frac{1}{\lim_{m\to\infty}(1 - \frac{1}{m})^m} \qquad (2.4)$$

But, $e = \frac{1}{e^{-1}}$

$$\therefore \quad e^{-1} = \lim_{m\to\infty}(1 - \frac{1}{m})^m \qquad (2.5)$$

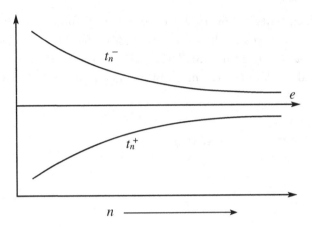

Figure 2.1: Convergence of sequence 1 and sequence 2

Table 2.3: Approaching towards e from both sides

n	$t_n^+ = \left(1 + \frac{1}{n}\right)^n$	$< e <$	$t_n^- = \left(1 - \frac{1}{n}\right)^{-n}$
10	2.593742	$< e <$	2.867971
100	2.704813	$< e <$	2.731999
1000	2.716923	$< e <$	2.719642
10000	2.718145	$< e <$	2.718417
100000	2.718268	$< e <$	2.718295

t_n^+ and t_n^- values are calculated on the basis of equation (2.3) and equation (2.4) respectively. In Table 2.3 we notice that t_n^+ and t_n^- come closer to each other as n increases. We may now draw two curves using t_n^+ and t_n^- values as shown in Figure 2.1.

So far we have come to know e as

1. The ultimate outcome of continuous compounding of Rs. 1 at the end of 1 year.

2. The sum of an infinite series — refer equation (2.3)

3. The asymptotic value of two curves drawn on the basis of equation (2.3) and equation (2.4) (Refer Figure 2.1).

2.5   Proof of convergence of two sequences of e

Jacob Bernoulli discovered the constant e in 1683 while studying the problem related to compound interest. He showed that the following two sequences converge.

Sequence 1: $t_n{}^+ = \left(1 + \frac{1}{n}\right)^n$, $n = 1, 2, 3, \cdots$

Sequence 1: $t_n{}^- = \left(1 - \frac{1}{n}\right)^{-n}$, $n = 1, 2, 3, \dots$

Table 2.4 shows values of $t_n{}^+$ and $t_n{}^-$ for different values of n.

Table 2.4: $t_n{}^+$ and $t_n{}^-$ for different values of n

n	1	2	3	4	5	6	7	8
$t_n{}^+$	2.000	2.250	2.370	2.441	2.448	2.552	2.594	2.704
$t_n{}^-$	-	4.000	3.375	3.161	3.052	2.986	2.868	2.732

Step A

$$1 + \frac{1}{n} = \frac{n+1}{n} = \frac{n+1}{(n+1)-1} = \frac{1}{1 - \frac{1}{n+1}} = \left(1 - \frac{1}{n+1}\right)^{-1}$$

$$\therefore \quad \left(1 + \frac{1}{n}\right)^{n+1} = \left(1 - \frac{1}{n+1}\right)^{-(n+1)}$$

$$\therefore \quad \left(1 + \frac{1}{n}\right)^n \left(1 + \frac{1}{n}\right) = t_{n+1}^-$$

$$\therefore \quad t_n^+ \left(1 + \frac{1}{n}\right) = t_{n+1}^-$$

$$\therefore \quad t_{n+1}^- - t_n^+ = \frac{t_n^+}{n} \tag{2.6}$$

$$\therefore \quad t_n^+ < t_{n+1}^- \tag{2.7}$$

Theorem of Geometric Series

$$\frac{a^{n+1} - b^{n+1}}{a - b} = a^n + a^{n-1}.b + a^{n-2}.b^2 + \cdots + a.b^{n+1} + b^n \tag{2.8}$$

[Product of denominator of L.H.S. and R.H.S. is
$$a^n(a-b) + a^{n-1}.b.(a-b) + \cdots + a.b^{n-1}.(a-b) + b^n.(a-b)$$
$$= a^{n+1} - a^n.b + a^n.b - \cdots + a.b^n + a.b^n - b^{n+1}$$
$$= a^{n+1} - b^{n+1}]$$

Step B

Let us now prove (i) $t_n^+ < t_{n+1}^+$ and (ii) $t_n^- > t_{n+1}^-$

Let us now consider equation (2.8) for a special case when $0 < b < a$

R.H.S. of equation (2.8) contains $(n+1)$ terms and the first term

is

a^n. For $b = a$ remaining n terms would also be equal to a^n.
Since $b < a$ each of these terms will be less than a^n, R.H.S. of
equation (2.8) $< (n+1).a^n$
$\therefore$ L.H.S. of equation (2.8) $< (n+1).a^n$

Again R.H.S. of equation (2.8) $> (n+1).b^n$
$\therefore$ L.H.S. of equation (2.8) $> (n+1).b^n$

So, we can write

$$(n+1).b^n < \frac{a^{n+1} - b^{n+1}}{a-b} < (n+1).a^n \qquad (2.9)$$

Proof of (i) $t_n^+ < t_{n+1}^+$

Let $a = 1 + \frac{1}{n}$ and $b = 1 + \frac{1}{n+1}$

From the R.H.S. of inequation (2.9) we can write

$$\frac{(1+\frac{1}{n})^{n+1} - (1+\frac{1}{n+1})^{n+1}}{(1+\frac{1}{n}) - (1+\frac{1}{n+1})} < (n+1)(1+\frac{1}{n})^n \qquad (2.10)$$

Simplifying the denominator of L.H.S. of inequation (2.10) we get
$$(1+\frac{1}{n}) - (1+\frac{1}{n+1}) = \frac{1}{n.(n+1)}$$

Multiplying inequation (2.10) by $\frac{1}{n.(n+1)}$

$$(1+\frac{1}{n})^{n+1} - (1+\frac{1}{n+1})^{n+1} < \frac{1}{n}.(1+\frac{1}{n})^n$$

$$\therefore \quad t_n^+ \left(1 + \frac{1}{n}\right) - t_{n+1}^+ < \frac{t_n^+}{n}$$

$$\therefore \quad t_n^+ < t_{n+1}^+ \tag{2.11}$$

Proof of (ii) $t_n^- < t_{n+1}^-$

Now we choose $a = 1 - \frac{1}{n+1}$ and $b = 1 - \frac{1}{n}$

L.H.S. of inequation (2.9) can now be rewritten as

$$(n+1).\left(1 - \frac{1}{n}\right)^n < \frac{\left(1 - \frac{1}{n+1}\right)^{n+1} - \left(1 - \frac{1}{n}\right)^{n+1}}{\left(1 - \frac{1}{n+1}\right) - \left(1 - \frac{1}{n}\right)} \tag{2.12}$$

Simplifying the denominator of R.H.S. of inequation (2.12) we get,

$$\left(1 - \frac{1}{n+1}\right) - \left(1 - \frac{1}{n}\right) = \frac{1}{n} - \frac{1}{n+1} = \frac{1}{n(n+1)}$$

$$\therefore \qquad \frac{1}{n}.\left(1 - \frac{1}{n}\right)^n < \left(1 - \frac{1}{n+1}\right)^{n+1} - \left(1 - \frac{1}{n}\right)^{n+1}$$

or, $\qquad \frac{1}{n}.\frac{1}{t_n^-} < \frac{1}{t_{n+1}^-} - \frac{1}{t_n^-}\left(1 - \frac{1}{n}\right)$

or, $\qquad \frac{1}{t_n^-} < \frac{1}{t_{n+1}^-}$

$$\therefore \quad t_n^- > t_{n+1}^- \tag{2.13}$$

STEP C: We will now show that by increasing the number n it is possible to make the difference $(t_{n+1}^- - t_n^+)$ as small as we please.

We may now ask "Is it possible to get the following expression and make ϵ as small as we please?"

$$(t_{n+1}^- - t_n^+) \leq \epsilon$$

From equation (2.13) we know that $t_{n+1}^- < t_n^-$

$$\therefore \textit{ for } n > 2 \quad t_{n+1}^- < t_2^-$$

From Table 2.4 we get $t_2^- = 4$

$$\therefore \textit{ for } n > 2 \quad t_{n+1}^- < 4$$

From equation (2.7) we know that $t_n^+ < t_{n+1}^-$

So, for $n > 2, \quad t_n^+ < 4$

From equation (2.6) we know,

$$t_{n+1}^- - t_n^+ = \frac{t_n^+}{n}$$

$$\therefore \ t_{n+1}^- - t_n^+ < \frac{4}{n} \text{ for } n > 2$$

By choosing larger and larger values of n we can have smaller and smaller values of $\frac{4}{n}$. This ensures that the sequences 1 and 2 converge to a single value.

Convergence of the two limits to a single value ensured that the converged value can be treated as a number. However, mathematicians proved that it cannot be a rational number. Later on it was found to be a non-algebraic (transcendental) type of irrational number like π. It took three centuries to reveal the real nature of this number after it appeared in disguise in the table of calculation of compound interest. The formula $(1 + \frac{1}{n})^n$ belongs to commercial arithmetic; but when some one starts dealing with $\lim_{n \to \infty}(1 + \frac{1}{n})^n$ he/she engages in abstract mathematics.

2.6   Proof of Irrationality of e

This proof, based on *Reductio ad absurdum* (or method of contradiction), was first given by Joseph Fourier in 1815.

To start with let us assume that e is rational. So we can write $e = \frac{p}{q}$ where p and q are integers.

Now, $e = 1 + (\frac{1}{1!}) + (\frac{1}{2!}) + (\frac{1}{3!}) + \cdots$ up to ∞.

$$\therefore \ (\frac{p}{q}) = 1 + (\frac{1}{1!}) + (\frac{1}{2!}) + (\frac{1}{3!}) + \cdots + (\frac{1}{q!}) + [\frac{1}{(q+1)!}] + [\frac{1}{(q+2)!}] + \cdots \infty$$

Multiplying both sides by $q!$ we get,

$$(\frac{p}{q}).q! = q! + (\frac{q!}{1!}) + (\frac{q!}{2!}) + (\frac{q!}{3!}) + \cdots + (\frac{q!}{q!}) + [\frac{q!}{(q+1)!}] + [\frac{q!}{(q+2)!}] + \cdots \infty$$

or, $p.(q-1)! - q! - (\frac{q!}{1!}) - (\frac{q!}{2!}) - (\frac{q!}{3!}) - \cdots - (\frac{q!}{q!}) = [\frac{q!}{(q+1)!}] + [\frac{q!}{(q+2)!}] + \cdots \infty$

We note that each term of the L.H.S. of the above equation is an integer. As such, L.H.S. finally yields an integer value.

$$\text{R.H.S.} = \frac{q!}{(q+1)!} + \frac{q!}{(q+2)!} + \cdots \infty < \frac{1}{(q+1)} + \frac{1}{(q+1)^2} + \frac{1}{(q+1)^3} \cdots \infty$$

$$\therefore \ \text{R.H.S.} < \frac{\frac{1}{(q+1)}}{(1-\frac{1}{q+1})}$$

$$\therefore \ \text{R.H.S.} < \frac{1}{q}$$

Since q is an integer R.H.S. is less than 1 and can not be an integer. However, L.H.S. is an integer. So there is a contradiction and our initial assumption that e is a rational number can not be true. Therefore, $e \neq (\frac{p}{q})$.

2.7 Function e^x

In our example of calculation of compound interest if B is allowed to pay back the money after 2 years the final amount will be e^2. It is because of the fact that after the end of one year Rs. 1 has become Rs. e; so after the end of next one year (i.e. second year) Rs. e becomes Rs. e^2. Using the same logic we can say that at end of three years Rs. 1 becomes e^3. So let us now formulate a general problem — "what would happen to the final amount if it is allowed to grow by continuous compounding for x years at an interest rate of 100% ?" We can also formulate a second general problem — "what would happen to the final amount if it is allowed to grow by continuous compounding for 1 year at an interest rate of x times 100%?". We will afterwards see that the result of both these problems will be same.

Case 1: $e^x = \lim_{n \to \infty}(1 + \frac{1}{n})^{nx}$

Expanding the function $(1 + \frac{1}{n})^{nx}$ we get

$$
\begin{aligned}
e^x &= \lim_{n \to \infty} [1 + nx.(\frac{1}{n}) + \frac{nx.(nx-1)}{2!}.(\frac{1}{n})^2 \\
&\quad + \frac{nx.(nx-1)(nx-2)}{3!}.(\frac{1}{n})^3 + \cdots \infty] \\
&= \lim_{n \to \infty} [1 + x + \frac{x.(x-\frac{1}{n})}{2!} + \frac{x.(x-\frac{1}{n}).(x-\frac{2}{n})}{3!} + \cdots \infty]
\end{aligned}
$$

$$
\therefore \quad e^x = 1 + x + \frac{x^2}{2!} + \frac{x^3}{3!} + \dots to \ \infty
$$

$$(2.14)$$

Case 2: $e^x = \lim_{n \to \infty}(1 + \frac{x}{n})^n$

Expanding the function $(1 + \frac{x}{n})^n$ we get

$$
e^x = \lim_{n \to \infty} [1 + n\frac{x}{n} + \frac{n(n-1)(\frac{x^2}{n^2})}{2!} + \frac{n.(n-1).(n-2)(\frac{x^3}{n^3})}{3!} + \cdots \infty]
$$

$$= \lim_{n \to \infty}[1 + x + \frac{(1-\frac{1}{n})x^2}{2!} + \frac{(1-\frac{1}{n})(1-\frac{2}{n})x^3}{3!} + \cdots \infty]$$

$$= 1 + x + \frac{x^2}{2!} + \frac{x^3}{3!} + \cdots \infty$$

Since we get same result for both **Case 1** and **Case 2** we may write

$$\therefore \quad e^x = \lim_{n \to \infty}(1+\frac{1}{n})^{nx} = \lim_{n \to \infty}(1+\frac{x}{n})^n = 1 + x + \frac{x^2}{2!} + \frac{x^3}{3!} + \cdots \infty$$

$$(2.15)$$

However, it is important to note that for any finite value of n, the expressions $(1 + \frac{x}{n})^n$ and $(1 + \frac{1}{n})^{nx}$ will produce different results for any finite value of x.

Table 2.5: A_n and B_n for different values of n

n	@200% interest after 1 year $A_n = (1+\frac{2}{n})^n$	@100% interest after 2 years $B_n = (1+\frac{1}{n})^{2n}$
1	3	4
10	6.1917	6.7274
100	7.2446	7.3160
1000	7.3743	7.3816
10000	7.3876	7.3883
100000	7.3889	7.3890

Let us now consider two concrete cases. In the first case the rate of interest is 200% and matured amount will be paid at the end of one year. The second column of the Table 2.5 shows the matured amount for this case for different values of n. In the second case the rate of interest is 100% and matured amount will be paid at the end of second year. The third column of the Table 2.5 shows the matured amount for this case for different values of n. We notice that for small values of n the matured amount of two cases are quite different. But as the n becomes larger and larger the difference keeps on decreasing. For very large values of n they become very close to e^2.

$e^2 = 7.389056099$ (as per Texas Instrument Calculator TI36X Pro)

Suppose u is the per unit annual interest rate (i.e., % annual interest rate divided by 100), v the number of years and $x = u.v$

$\therefore \quad e^x = \lim_{n \to \infty}(1 + \frac{u}{n})^{nv} = 1 + x + \frac{x^2}{2!} + \frac{x^3}{3!} + \dots \text{to } \infty$

So *the exponent x in e^x is the product of number of years and per unit annual interest rate.*

In equation (2.15) if we use mx as the exponent of e we get the following equation.

$$\therefore \quad e^{mx} = 1 + mx + \frac{m^2 x^2}{2!} + \frac{m^3 x^3}{3!} + \cdots \infty \tag{2.16}$$

Figure 2.2 shows the variation of e^{mx} for different values of m.

2.8 Interesting features of exponential functions e^x and e^{mx}

It is interesting to note if we differentiate e^x it remains same [refer equation (2.15)] since the 1st term 1 of the series becomes zero and each successive term after differentiation becomes equal to its predecessor of the original series. So the 2nd term x after differentiation becomes 1, the 3rd term $\frac{x^2}{2!}$ after differentiation becomes x and so on and infinite series remains same. This is a remarkable feature of e^x.

$$\frac{de^x}{dx} = e^x \tag{2.17}$$

$$\therefore \quad \int e^x . dx = e^x + C \tag{2.18}$$

where C is the integration constant. If $c = 0$, $\int e^x . dx = e^x$.

If we now differentiate e^{mx} we get

$$\begin{aligned} \frac{de^{mx}}{dx} &= m + m^2 x + m^3 \frac{x^2}{2!} + m^4 \frac{x^3}{3!} + \cdots \infty \\ &= m(1 + mx + m^2 \frac{x^2}{2!} + m^3 \frac{x^3}{3!} + \cdots \infty) = m.e^{mx} \end{aligned} \tag{2.19}$$

So we note that the derivative of e^{mx} is proportional to itself and the proportionality constant is m.

If we further differentiate L.H.S. of equation (2.19) we get

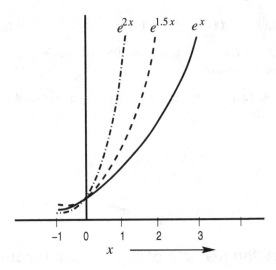

Figure 2.2: e^{mx} for different values of m

$$\frac{d^2 e^{mx}}{dx^2} = m^2.e^{mx} \tag{2.20}$$

In a similar way we find

$$\frac{d^2 e^{-mx}}{dx^2} = m^2.e^{-mx} \tag{2.21}$$

Observing equation (2.20) and equation (2.21) we conclude that if the second derivative of a function is proportional to itself the function may be $A.e^{mx}$ and/or $B.e^{-mx}$ where A and B are constants. We will utilise this observation later in section 2.20 in this chapter while solving a second order differential equation.

It is worthwhile to note that

$$\int e^{mx}\, dx = \frac{e^{mx}}{m} + C \tag{2.22}$$

Here C is the integration constant.

2.9 Story of snails or an informal explanation of e^1 vis-a-vis e^i

Long back, when I was a child, I read in a fairy tale book that a snail's mental world is quite different from ours. We may think they are lazy creatures; but they do not think so. What is 1 second to us appears to be almost a year to them. Our 1 meter may be their 1 km. They perceive the world in a different way. So I thought it would be appropriate to involve snails when we need to deal with infinitesimals in mathematics.

A story goes like this — once upon a time six snails (S_1, S_2, S_3) and (S_4, S_5, S_6) huddled together at the crossing of two roads. This is shown as point C in Figure 2.3. They form two groups — $(S_1$ *to* $S_3)$–Group A and $(S_4$ *to* $S_6)$–Group B. Point C was 1 km Eastward from a tree shown as point O. The two groups decided to travel in different directions. Members of Group A decided to move further in the eastward direction along the road OC. But members of Group B decided to move towards north through a road which runs perpendicular to the earlier road at C.

2.9.1 e as a distance travelled along a road

Snail S_1 decided to travel 1 km in the next year and reach $D^A_{1,1}$ at the end of the year where $OD^A_{1,1}$=2km.

Snail S_2 decided to walk at same speed that of S_1 for the first three months of the year and reached $D^A_{4,1}$ where $OD^A_{4,1} = 1.25000$ km. It then decided to walk one fourth of $OD^A_{4,1}$ i.e., 0.3125 km during the second quarter (next 3 months) and reached $D^A_{4,2}$ where $OD^A_{4,2} = 1.56250 (= 1.25000 + 0.31250)$ km. He decided to cover one fourth of $OD^A_{4,2}$ during the third quarter and reached $D^A_{4,3}$ where $OD^A_{4,3} = 1.95312$ km. He decided to cover one fourth of $OD^A_{4,3}$ during the fourth and last quarter and reached $D^A_{4,4}$ where $OD^A_{4,4} = 2.44141$ km. Thus S_2 travelled greater distance than S_1.

We may, however, use the following mathematical formula for computational purpose.

Distance travelled after k^{th} interval

$$OD^A_{n,k} = (1 + \frac{1}{n})^k \qquad (2.23)$$

where $k = 1, 2, \cdots, n$

Here n is the total number of intervals into which one year has been divided.

Distance travelled by S_2 at the end of 1st interval ($k = 1$)

$$OD^A_{4,1} = (1 + \tfrac{1}{4})^1 = 1.25000$$

Distance travelled by S_2 at the end of 2nd interval ($k = 2$)

$$OD^A_{4,2} = (1 + \tfrac{1}{4})^2 = 1.56250$$

Distance travelled by S_2 at the end of 3rd interval ($k = 3$)

$$OD^A_{4,3} = (1 + \tfrac{1}{4})^3 = 1.95312$$

Distance travelled by S_2 at the end of 4th interval ($k = 4$)

$$OD^A_{4,4} = (1 + \tfrac{1}{4})^4 = 2.44141$$

Snail S_3 divided one year interval into n number of intervals where n is infinitely large. The distance travelled in any interval is $\frac{1}{n}$ times the distance of its position from the point O at the end of previous interval. As such, at the end of one year S_3 will be at a distance $(1 + \frac{1}{n})^n$ from the point O. Since n is infinitely large we may write it in the following manner.

$$OD^A_{\infty,\infty} = \lim_{n \to \infty} (1 + \frac{1}{n})^n = e \; (\approx 2.71828).$$

2.9.2 e^i as a distance travelled in a field

In equation (2.15) if we replace x by i we get $e^i = \lim_{n\to\infty}(1 + \frac{i}{n})^n$. We will now try to find a physical interpretation for the expression $(1 + \frac{i}{n})^n$ both for finite and infinite values of n.

When $n = 1$, we get $(1 + i)$. If any one moves 1 km north ward from C (Refer Figure 2.4) he will be at a distance i km from C and $(1 + i)$km from O. If we divide 1 year into n number of equal time intervals then any body walking at a speed of 1km per year towards north will cover a distance $\frac{i}{n}$ km in 1st interval (i.e., $\frac{1}{n}$ year) from C. He will be, however, at $D^B_{n,1}$ which is at a distance $(1 + \frac{i}{n})$ km from O. In the second interval if he travels $\frac{i}{n}$ times the earlier distance (i denotes that movement takes place in perpendicular direction) he will

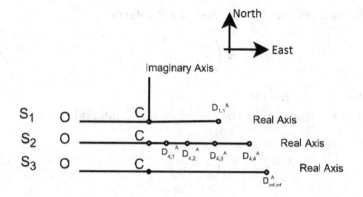

Figure 2.3: e as a distance travelled along a road

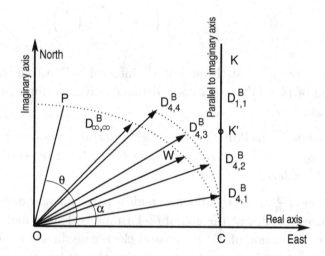

Figure 2.4: e^i as a distance travelled in a field

travel $\frac{i}{n}(1 + \frac{i}{n})$ km in the second interval. So at the end of second interval he will be at a distance $[(1 + \frac{i}{n}) + \frac{i}{n}(1 + \frac{i}{n})] = (1 + \frac{i}{n})^2$ km from O. After the completion of all the n intervals (1 year) he will be at a distance $(1 + \frac{i}{n})^n$ km from O.

For Group B we may use a similar equation as equation (2.23); however, we need to replace $\frac{1}{n}$ by $\frac{i}{n}$.

Distance travelled at the end of kth interval

$$OD^B_{n,k} = \left(1 + \frac{i}{n}\right)^k \tag{2.24}$$

where $k = 1, 2, \cdots n$

Distance travelled at the end of $(k-1)^{th}$ interval

$$OD^B_{n,k-1} = \left(1 + \frac{i}{n}\right)^{k-1} \tag{2.25}$$

Distance travelled during the kth interval is

$$\left(1 + \frac{i}{n}\right)^k - \left(1 + \frac{i}{n}\right)^{k-1} = \left(1 + \frac{i}{n}\right)^{k-1}\left[\left(1 + \frac{i}{n}\right) - 1\right]$$

$$= \left(1 + \frac{i}{n}\right)^{k-1}\left(\frac{i}{n}\right) \tag{2.26}$$

Distance travelled at the end of kth interval = Distance travelled at the end of $(k-1)^{th}$ interval + Distance travelled during the kth interval

$\therefore$ *Distance travelled at the end of kth interval = Distance travelled at the end of $(k-1)^{th}$ interval + $\frac{i}{n}$ times the distance travelled during the $(k-1)^{th}$ interval*

Addition of $\frac{i}{n}$ times the distance adds a perpendicular distance to the distance travelled at the end of (k-1)th interval since multiplying by i causes a rotation of $90°$ in the anti-clock wise direction. As such during each interval a snail changes its direction of travel.

Snail S_4 decided to travel 1 km towards north in the next year and reach $D^B_{1,1}$ at the end of the year where $OD^B_{1,1} = (1 + i1) \approx 1.41 \angle 45°$.

Snail S_5 decided to walk at same speed and in the same direction as that of S_4 for the first three months of the year and reached $D^B_{4,1}$ where $OD^B_{4,1} = (1 + i0.25)$ km $\approx 1.03 \angle 14.04°$. He then decided to walk $\frac{i}{4}$ times (**not** $\frac{1}{4}$) of $OD^B_{4,1}$ [= (-0.0625 + i0.25)km] during the second quarter (next 3 months) and reached $D^B_{4,2}$ where $OD^B_{4,2} = (0.9375 + i0.5)$km $\approx 1.06 \angle 28.07°$. He decided to cover $\frac{i}{4}$ times of $OD^B_{4,2}$ during the third quarter and reached $D^B_{4,3}$ where $OD^B_{4,3} = (0.9375 + i0.5) + 0.25i(0.9375 +$

$i0.5) = 0.8125 + i0.734375 \approx 1.09\angle42.11°$. He decided to cover $\frac{i}{4}$ times of $OD_{4,3}^B$ during the fourth and last quarter and reached $D_{4,4}^B$ where $OD_{4,4}^B = (0.8125 + i0.734375) + 0.25i(0.8125 + i0.734375) = 0.62890625 + i0.9375 \approx 1.13\angle56.14°$. So S_5 travels a little bit more distance at each quarters and also changes direction towards west.

It is interesting to note the following points.

1. The magnitude of $OD_{4,4}^B$ is less than that of $OD_{1,1}^B$ and it has come closer to 1.

2. The angle of $OD_{4,4}^B$ has come closer to 57.29° (1 radian).

3. When $n > 4$ the point reached after 1st interval lies below $D_{4,1}^A$. Larger the n lower will be this point and it will come nearer to point C.

Snail S_6 divided one year into infinitesimal number of intervals i.e. $n = \infty$. The point reached after the 1st interval will be infinitesimally close to C. The distance travelled by S_6 at the end of kth interval is $(1 + \frac{i}{n})^k$.

The expression inside the parenthesis $(1 + \frac{i}{n})$is a complex quantity with a real part of 1 and imaginary part of $\frac{1}{n}$. As such, the magnitude of this complex quantity $\sqrt{(1 + \frac{1}{n^2})}$ and its angle will be $\tan^{-1}\frac{1}{n}$. When n is very large and approaches ∞ we can neglect $\frac{1}{n^2}$ and we can assume angle $\tan^{-1}\frac{1}{n} = \frac{1}{n}$. As such, the magnitude of the complex quantity $(1 + \frac{i}{n})^k$ is equal to 1 and its angle is $\frac{k}{n}$ radian. As k changes from 1 to n, the angle changes from 0 radian to 1 radian; but the magnitude $OD_{k,k}^B$ remains same $(= 1)$. So S_6 moves over the circle of unit radius starting from C (angle is zero radian) and ending at $D_{\infty,\infty}^B$ (angle is 1 radian).

$$\therefore \quad OD_{\infty,\infty}^B = \lim_{n\to\infty}(1 + \frac{i}{n})^n = (1\angle\frac{1}{n})^n = (1)^n\angle n.\frac{1}{n} = 1\angle1 rad = 1\angle57.29°$$

2.10 General function $e^{i\theta} = \lim_{n\to\infty}(1 + \frac{i\theta}{n})^n$

The expression inside the parenthesis represents a complex variable with a magnitude of $\sqrt{1^2 + (\frac{\theta}{n})^2}$ and since n is very large it will have a value 1. The angle of the variable will be $\tan^{-1}\frac{\theta}{n} \approx \frac{\theta}{n}$ for small values of $\frac{\theta}{n}$.

$$\therefore \ \lim_{n\to\infty}(1+\tfrac{i\theta}{n})^n = 1^n \angle n.\tfrac{\theta}{n} = 1\angle\theta$$

$$\therefore \ e^{i\theta} = \cos\theta + i\ \sin\theta \qquad (2.27)$$

Pictorially $e^{i\theta} = 1\angle\theta$ is shown as the line OP in Figure 2.6.

It will now be worthwhile to attach this limit to a thought experiment and read its implication in a practical manner.

A horizontal line OC has a length 1 km. A tree is located at C and is shown as a vertical line CK where $CK = \theta$ km. With O as centre and OC as radius a circle is drawn in the vertical plane. In real world this circle may represent a ring. Since it is an unity circle, for any point W on it the length of the arc $CW(=\alpha)$ is equal to $\angle COW(\angle COW = $ Length of arc $CW \div$ Radius $= \tfrac{\alpha}{1} = \alpha)$. Let us assume that the trunk CK of the tree is highly elastic and flexible. As such if a rope is tied at K and a person sitting at O starts pulling the rope towards the centre of the circle O, finally the length will align with a circular sector CP where $CP = CK = \theta$. However, this process of alignment takes place gradually and the alignment starts from the lower portion of the tree and ends when the top most point K moves to P which is located on the circumference of the unit circle. As such, a point K' aligns with the circle earlier than point K.

So the limit $\lim_{n\to\infty}(1+\tfrac{i\theta}{n})^n$ *mathematically narrates the final outcome of the process of continuous bending and aligning of a line over a circle. At the end a straight line gets converted to an arc of same length and generates an angle of same radian value since the associated circle has unit radius.*

2.11 Infinite series representation of $\cos\theta$ and $\sin\theta$

Infinite series representation of $e^{i\theta}$ is given below.

$$e^{i\theta} = 1+i\theta+\tfrac{(i\theta)^2}{2!}+\tfrac{(i\theta)^3}{3!}+\tfrac{(i\theta)^4}{4!}+\tfrac{(i\theta)^5}{5!}+\tfrac{(i\theta)^6}{6!}+\tfrac{(i\theta)^7}{7!}.....to\ \infty$$

$$= 1+i\theta+\tfrac{i^2\theta^2}{2!}+\tfrac{i^3\theta^3}{3!}+\tfrac{i^4\theta^4}{4!}+\tfrac{i^5\theta^5}{5!}+\tfrac{i^6\theta^6}{6!}.+\tfrac{i^7\theta^7}{7!}.....to\ \infty$$

We know that $i=\sqrt{-1}$; $i^2=-1$, $i^3=-i$; $i^4=1$; $i^5=i$; $i^6=-1$; $i^7=-i$

$$\therefore \ e^{i\theta} = (1 - \frac{\theta^2}{2!} + \frac{\theta^4}{4!} - \frac{\theta^6}{6!}....) + i(\theta - \frac{\theta^3}{3!} + \frac{\theta^5}{5!} - \frac{\theta^7}{7!}....) \tag{2.28}$$

equation (2.20)

Equating real parts of equation (2.27) and equation (2.28) we get

$$\cos\theta = (1 - \frac{\theta^2}{2!} + \frac{\theta^4}{4!} - \frac{\theta^6}{6!} + \cdots \ \infty) \tag{2.29}$$

Equating imaginary parts of equation (2.27) and equation (2.28) we get

$$\sin\theta = (\theta - \frac{\theta^3}{3!} + \frac{\theta^5}{5!} - \frac{\theta^7}{7!} + \cdots \ \infty) \tag{2.30}$$

If we raise the power (exponent) of $e^{i\theta}$ by n we get $e^{in\theta}$. Replacing θ by $n\theta$ in equation (2.27) we now write

$$e^{in\theta} = \cos(n\theta) + i\,\sin(n\theta) \tag{2.31}$$

The above equation is known as De Moivre's formula.

When $\theta = \omega t$ (ω is angular velocity and t is time), $e^{i\theta}$ becomes a rotating vector of unit length since θ increases with time. The expression $re^{i\theta}$ represents a rotating vector of length r. The loci of these vectors will be circles of radius 1 and r respectively (refer Figure 2.5).

2.12   Raising the power of e by complex angle $(\alpha + i\theta)$

Let us understand it in terms of the distance travelled in a field due to different types of movement.

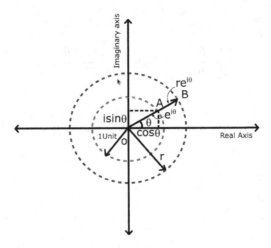

Figure 2.5: Rotating Vectors $e^{i\theta}$ and $r.e^{i\theta}$

We know that

$$e^{\alpha+i\theta} = e^{\alpha}.e^{i\theta} = r.e^{i\theta} = r(\cos\theta + i\sin\theta) = r\cos\theta + ir\sin\theta$$

where $r = e^{\alpha}$.

In Figure 2.5 we find a planar vector OB which can be symbolically expressed as $r.e^{i\theta}$. We may,however, replace r by e^{α} (e.g., when $r = 3.9$, $\alpha = 1.36$ since $3.9 = e^{1.36}$).

At this point we may introduce the concept of complex angle by treating the exponent $(\alpha + i\theta)$ of e as the angle. Earlier we have seen that the imaginary part of the exponent (i.e. θ) represents the angle of a planar vector. From now onwards we may also treat the real part of the exponent (i.e., α) also as an angle and it assumes different values when the magnitude of a vector changes. In Chapter 4 we will see that if the imaginary part (θ) of the complex angle is a circular angle the real part (α) is a hyperbolic angle.

At this point it is worth noting that a point in a complex plain can be represented in following forms.

- Rectangular Form $(a + ib)$

- Polar Form $(r\angle\theta)$ where $r = \sqrt{a^2 + b^2}$ and $\theta = \tan^{-1}\frac{b}{a}$

- Exponential Form $e^{\alpha+i\theta}$ where $e^{\alpha} = r$

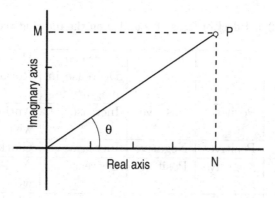

Figure 2.6: Point P in the complex plane

It is, however, necessary to look into the matter in a more critical manner. Let us consider a point P (refer Figure 2.6) which is expressed as $(4 + i3)$ in the rectangular form. The corresponding polar form associated with the vector OP can be written as $5\angle 0.64$ rad [$5 = \sqrt{4^2 + 3^2}$ and $\tan^{-1}\frac{3}{4} = 0.64$ rad.]. If the vector OP now undergoes one, two or any integral number of complete 2π rotation the rectangular representation remains unchanged.

Since $e^{1.61} \approx 5$, we may write

$$4 + i3 = 5\angle(0.64 + 2\pi k) \text{ rad} = e^{1.61 + i(0.64 + 2\pi k)} \text{ where } k = 0, 1, 2 \cdots \infty$$

Since k can have infinite number of values, a particular rectangular notation will have multiple and infinite number of polar and exponential representations. As such, logarithm of a complex number may have infinite number of solutions. This point will be discussed again in Chapter 3.

2.13 Rotating vector and concept of complex frequency

Now let us consider a planar vector OA [refer Figure 2.8] which keeps on rotating in the anti-clockwise direction over a plane. During its rotation the vector undergoes an increase in the imaginary part of the angle θ and an increase in magnitude (and hence a change in the real part of the angle α). The locus of the movement of the tip of the

Table 2.6: Effect of signs of σ and ω on the rotating vector

Figure number	σ	ω	Increase/ Decrease in magnitude	Direction of Rotation
2.8	Positive	Positive	Increase	Anti-clock wise
2.9	Positive	Negative	Increase	Clock wise
2.11	Negative	Positive	Decrease	Anti-clock wise
2.12	Negative	Negative	Decrease	Clock wise

vector is shown as a spiral. If the rates at which α and θ increase are σ and ω respectively we get $\alpha = \sigma t$ and $\theta = \omega t$ where t is the elapsed time. So the vector can be expressed as $e^{\sigma t}.e^{i\omega t} = e^{\sigma t + i\omega t} = e^{(\sigma + i\omega)t}$. If rotation of the vector is not associated with change in magnitude $\sigma = 0$ and the locus of movement of the tip of the vector will be a circle. The part of the exponent inside the parenthesis $(\sigma + i\omega)$ is the rate at which the complex angle changes. We call $(\sigma + i\omega)$ an expression for complex frequency. Depending on the sign of $(\sigma$ and $i\omega)$ we may encounter four types of movement of a vector as narrated in Table 2.6 and portrayed in Figure 2.8, Figure 2.9, Figure 2.11 and

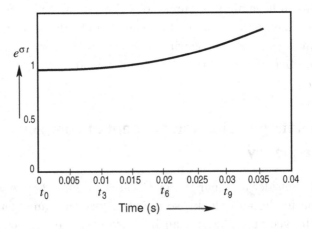

Figure 2.7: Graph of $e^{\sigma t}$

Figure 2.12. Figure 2.7 shows how $e^{\sigma t}$ changes with increase in t. Figure 2.10 shows how $e^{-\sigma t}$ changes with increase in t.

2.14 $\cos\theta$ and $\sin\theta$ in terms of exponential functions

In Figure 2.13 Length of OM = Length of ON = Length of OP = 1 unit

$$OM = e^{i\theta} = \cos\theta + i\ \sin\theta = OA + i\ OD \qquad (2.32)$$

$$ON = e^{-i\theta} = \cos\theta - i\ \sin\theta = OA - i\ OF \qquad (2.33)$$

Adding equation (2.32) and equation (2.33) we get
$$e^{i\theta} + e^{-i\theta} = 2OA + i(OD - OF) = 2OA = OB = 2\cos\theta$$

$$\therefore \quad \cos\theta = \frac{e^{i\theta} + e^{-i\theta}}{2} \qquad (2.34)$$

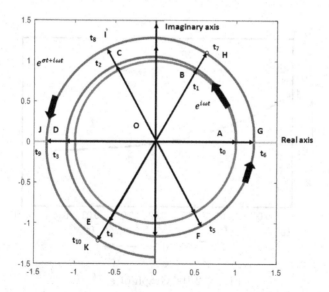

Figure 2.8: Graph of $e^{\sigma t + i\omega t}$ for +ve σ and +ve ω

Let us now assume that the point M starts moving in the anti-clock wise direction over a unit circle producing a time varrying θ where $\theta = \omega t$ (ω is angular speed and t is time) and the point N moves over the same unit circle in the clock wise direction. The combined effect of

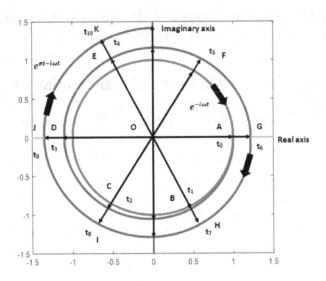

Figure 2.9: Graph of $e^{\sigma t + i\omega t}$ for +ve σ and -ve ω

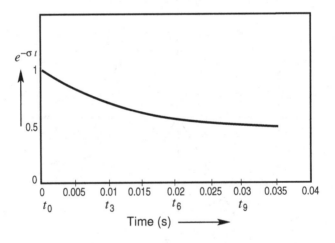

Figure 2.10: Graph of $e^{-\sigma t}$

two circular movements in opposite directions result in an oscillatory movement along the real axis. In equation (2.34), $\cos\theta$ represents an oscillatory movement along real axis and $\frac{e^{i\theta}}{2}$ and $\frac{e^{-i\theta}}{2}$ represent two oppositely directed rotational movement.

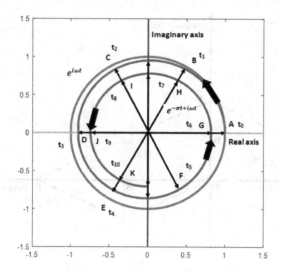

Figure 2.11: Graph of $e^{\sigma t + i\omega t}$ for -ve σ and +ve ω

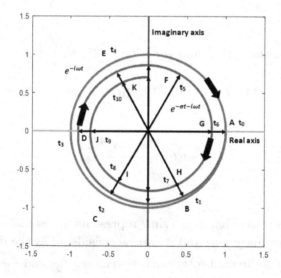

Figure 2.12: Graph of $e^{\sigma t + i\omega t}$ for -ve σ and -ve ω

Again if we refer to Figure 2.13 we find

$$OP = -e^{-i\theta} = -[\cos-\theta + i\ \sin-\theta] = -\cos\theta + i\sin\theta = -OC + iOD \tag{2.35}$$

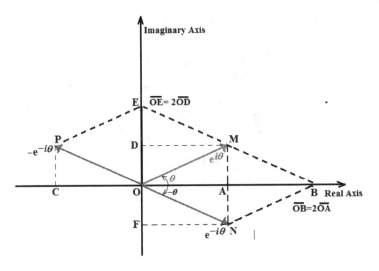

Figure 2.13: $e^{i\theta}$ and $e^{-i\theta}$ as planar vectors

Adding equation (2.32) and equation (2.35) we get

$$e^{i\theta} - e^{-i\theta} = (OA - OC) + 2iOD = 2iOD = iOE = i.2\sin\theta$$

$$\therefore \quad \sin\theta = \frac{e^{i\theta} - e^{-i\theta}}{2i} \tag{2.36}$$

When θ keeps on changing $i\sin\theta$ represents an oscillatory movement along imaginary axis and it can be considered as the difference of two oppositely directed rotational movements $\frac{e^{i\theta}}{2}$ and $\frac{e^{-i\theta}}{2}$.

(Electrical engineers come across a phenomenon in a standstill single phase induction motor which can be mathematically described by the equation (2.34) or equation (2.36). The stator winding of the motor produces a pulsating magnetic field which can be conceived as the sum of two rotating magnetic fields of equal magnitude which are rotating in opposite directions.)

2.15 Obtaining an ellipse as the resultant of two rotating planar vectors rotating in opposite direction

Difference of two vectors r_1 and r_2 rotating in opposite directions can be mathematically described by the following equation.

$$
\begin{aligned}
r_1 e^{i\theta} - r_2 e^{-i\theta} &= r_1 \cos\theta + i r_1 \sin\theta - r_2 \cos\theta + i r_2 \sin\theta \\
&= (r_1 - r_2)\cos\theta + i(r_1 + r_2)\sin\theta \\
&= a\cos\theta + ib\sin\theta
\end{aligned}
$$

where $a = (r_1 - r_2)$ and $b = (r_1 + r_2)$

Let, $x = a\cos\theta$ and $y = b\sin\theta$

$\therefore$ $\frac{x}{a} = \cos\theta$ and $\frac{y}{b} = \sin\theta$

$$
\therefore \quad \frac{x^2}{a^2} + \frac{y^2}{b^2} = 1 \tag{2.37}
$$

Equation (2.37) is the equation of an ellipse.

So we may say that if two vectors of unequal lengths start rotating in opposite directions the length of the resultant vector keeps on changing from the difference of their lengths $(r_1 - r_2)$ to the sum of their lengths $(r_1 + r_2)$ and the tip of the resultant vector lies on an ellipse (refer Figure 2.14).

(Electrical engineers come across a similar phenomenon in a rotating single phase induction motor. In such a case $r_1 e^{i\theta}$ represents a forward magnetic field and $r_2 e^{-i\theta}$ represents a backward ward magnetic field. For a single phase induction motor $r_2 < r_1$. The resultant rotating magnetic field follows an elliptic path).

2.16 Problems related to division of a number

Suppose we divide a number A into y equal parts each of which is equal to x.

$$
\therefore \quad \frac{A}{y} = x
$$

Now the product of y equal parts will be

$$
x.x.x \cdots y \text{ times} = x^y = x^{\frac{A}{x}} = z \text{ (say)}
$$

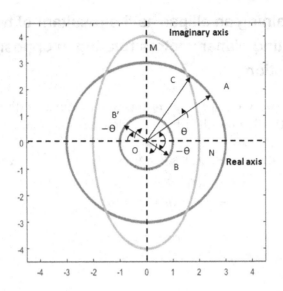

Figure 2.14: An ellipse as the resultant of two rotating vectors

Let us now find the x for which z will be maximum.
Taking log (base e) we get

$$\ln z = \frac{A}{x} . \ln x \qquad (2.38)$$

Differentiating equation (2.38) we get

$$\frac{1}{z} . \frac{dz}{dx} = -\frac{A}{x^2} . \ln x + \frac{A}{x} . \frac{1}{x}$$

$$\text{or,} \quad \frac{dz}{dx} = -\frac{zA}{x^2} . (1 - \ln x)$$

For z to be maximum, $\frac{dz}{dx} = 0$

$$\therefore \ \ln x = 1$$

$$\therefore \ x = e$$

2.16.1   Maximisation of the function $x^{\frac{1}{x}}$

If we now take $A = 1$, it becomes a problem of maximisation of the function $x^{\frac{1}{x}}$. This is known as *Steiner's Problem*.

Table 2.7: $x^{\frac{1}{x}}$ for different values of x

x	$x^{\frac{1}{x}}$
0.1	1×10^{-10}
1.0	1.000
2.0	1.414
2.5	1.442
e	**1.445**
3.0	1.442
3.5	1.430
.	.
.	.
10	1.259
100	1.047
1000	1.007
10000	1.001

In Table 2.16.1 we note that as x starts increasing from zero, $x^{\frac{1}{x}}$ keeps on increasing till it reaches the value 1.445 (approximately) at $x = e$. Further increase in the value of x causes decrease of the value of $x^{\frac{1}{x}}$ and it comes closer and closer to 1. *It is interesting to note that the function* $x^{\frac{1}{x}}$ *achieves it's maximum value at* $x = e$.

2.16.2 Dividing a number into several equal parts such that their product is maximum

Suppose we need to divide 18 into several equal parts such that their product will be maximum. Theoretically this product will be maximum if we divide 18 by e. Now $18 \div e \approx 6.62183$. The nearest integer is 7. Now, $18 \div 7 \approx 2.5714$. The product of seven 2.5714 is $(2.5714)^7 \approx 743.3399$. If we divide 18 into 6 equal parts each part will have a value 3. So the product of six equal parts is $3^6 = 729$. Again if we divide 18 into 8 equal parts each part will have a value 2.25. So the product of eight equal parts is $2.25^8 \approx 656.8408$. So when we divide 18 into integer number of equal parts the product of all the equal parts will be maximum when we divide it into seven equal parts.

2.17 Minimisation of the function x^x

Let $z = x^x$

$$\therefore \quad \ln z = x.\ln x \qquad (2.39)$$

Differentiating equation (2.39) and equating it to zero we get,

$$\frac{d\ln z}{dx} = \ln x + x.\frac{1}{x} = 0$$

$$\text{or,} \quad \ln x = -1$$

$$\text{or,} \quad x = e^{-1}$$

In Table 2.8 we note that as x starts decreasing from 1 towards zero, x^x keeps on decreasing till it reaches the value 0.6922 (approximately) at $x = e^{-1}$. Further decrease in the value of x brings x^x closer and closer to 1. *It is interesting to note that x^x attains minimum value at $x = e^{-1}$*

Table 2.8: x^x for different values of x

x	x^x
1	1
0.9	0.9095
0.8	0.8365
0.7	0.7790
0.6	0.7360
0.5	0.7071
0.4	0.6931
0.367 ($=e^{-1}$)	**0.6922**
0.3	0.6968
0.2	0.7278
0.1	0.7943
0.01	0.9550
0.001	0.9931
0.0001	0.9991
0.00001	0.9999

2.18 Computation of i^i

We know that $e^{i\theta} = \cos\theta + i\sin\theta$

When $\theta = \frac{\pi}{2}$, $e^{i\frac{\pi}{2}} = \cos\frac{\pi}{2} + i\sin\frac{\pi}{2}$

or, $e^{i\frac{\pi}{2}} = 0 + i.1 = i$

or, $i^i = (e^{i\frac{\pi}{2}})^i = e^{-\frac{\pi}{2}}$ *(a real number !!)*

2.19 Examples of 1st order differential equation

Uninhibited Growth

If a population can grow without any restriction its rate of change becomes proportional to its present population. Let $y(t)$ is the population at any instant of time and t denotes time.

$$\therefore \quad \frac{dy(t)}{dt} \propto y(t)$$

$$\text{or,} \quad \frac{dy(t)}{dt} = my(t) \tag{2.40}$$

Since we know that the rate of change of an exponential function is proportional to the function we can assume $y(t) = Ke^{mt}$ where K is a constant. Initially $t = 0$, $y(0) = K$.

$$\therefore \quad y(t) = y(0).e^{mt} \tag{2.41}$$

Heating (or cooling) of a body by its surrounding medium

If a metallic body is first cooled at a temperature T_1 and then it is placed in a large room with a higher temperature T_2 where $T_2 > T_1$ we know that finally the metallic body will attain a temperature T_2. However, during the intermediate period its temperature will keep on changing from T_1 to T_2 and the rate of change of the temperature difference between the surrounding medium and the metallic body at any instant will be proportional to this temperature difference at that instant.

Let $T(t)$ be the temperature of the metallic body at an instant t.

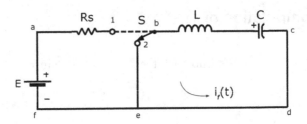

Figure 2.15: Circuit Diagram related to 2.20

$$\therefore \quad \frac{d[T_2 - T(t)]}{dt} \quad \propto \quad [T_2 - T(t)]$$

$$\text{or,} \quad \frac{d[T_2 - T(t)]}{dt} \quad = \quad m.[T_2 - T(t)]$$

$$\therefore \quad [T_2 - T(t)] \quad = \quad K.e^{-mt} \tag{2.42}$$

Initially at $t = 0$, $T(t) = T(0) = T_1$ $\therefore$ $K = T_2 - T_1$

2.20 Example of 2nd order differential equation

In the circuit shown in Figure 2.15 the switch S is initially at position 1. Since the capacitor C does not allow any direct current to flow throw it the battery will not supply any current and no current will flow through the circuit (a-b-c-d-e-f). However, the capacitor C will remain charged and the voltage across its terminals will be E. The charge stored by the capacitor C under this condition will be $q(0) = \frac{E}{C}$. If the switch is now placed at position 2, capacitor C will start discharging through the path (b-e-d-c) and a current will flow through this path. Since the capacitor starts discharging it keeps on supplying charge to flow through the circuit.

Let the charge supplied at an instant t be $q(t)$. Therefore, the current flowing through the circuit at t will be $i_f(t) = \frac{dq(t)}{dt}$. If we apply Kirchoff's Voltage Law in the loop(b-e-d-c) we get,

$$L.\frac{di_f(t)}{dt} + \frac{q(t)}{C} = 0$$

or, $$L.\frac{d^2q(t)}{dt^2} + \frac{q(t)}{C} = 0$$

or, $$\frac{d^2q(t)}{dt^2} = -\frac{q(t)}{LC} \qquad (2.43)$$

Let $\omega_0 = \frac{1}{\sqrt{LC}}$ $\therefore$ $i.\omega_0 = \frac{i}{\sqrt{LC}}$

Now, $-\frac{1}{LC} = (\frac{i}{\sqrt{LC}})^2 = (i.\omega_0)^2$

Replacing $-\frac{1}{LC}$ in equation (2.43) we get

$$\frac{d^2q(t)}{dt^2} = (i.\omega_0)^2.q(t) \qquad (2.44)$$

Equation (2.44) is a second order differential equation. We may now consider the following equation as the solution of equation (2.44) (refer equation (2.20) and equation (2.21)).

$$q(t) = A.e^{i\omega_0 t} + B.e^{-i\omega_0 t} \qquad (2.45)$$

In the above equation A and B are constants and we may find the values of these constants from initial conditions.

Now, $i_f(t) = \dfrac{dq(t)}{dt} = i\omega_0(A.e^{i\omega_0 t} - B.e^{-i\omega_0 t}) \qquad (2.46)$

Immediately after placing the switch on position 2(we start counting time from this instant) no current will flow through the circuit (b-e-d-c-b) i.e., $i_f(0) = 0$. This can be explained in the following manner. So long the switch S was in position 1 no current was flowing through L and C. Since inductance L does not allow any abrupt change in current through it no current will flow initially when switch S is placed in position 2.

In equation (2.46) if we put $t = 0$ we get,

$$i_f(0) = 0 = i\omega_0(A.1 - B.1) = i\omega_0(A - B)$$

$$\therefore \quad A = B$$

In equation (2.45) if we put $t = 0$ and $A = B$ we get,

$$q(0) = A.1 + A.1 = 2A$$

$$\therefore \quad A = \frac{q(0)}{2} = \frac{E}{2C}$$

Replacing A and B in equation (2.45) we get,

$$q(t) = \frac{q(0)}{2}.(e^{i\omega_0 t} + e^{-i\omega_0 t}) = q(0)\cos\omega_0 t = \frac{E}{C}\cos\omega_0 t \quad (2.47)$$

$$\therefore \quad i_f(t) = \frac{dq(t)}{dt} = -\frac{\omega_0 E}{C}\sin\omega_0 t \quad (2.48)$$

2.21 Miscellaneous examples

2.21.1 Radioactive Disintegration

Radioactive materials (e.g., carbon-14, plutonium-241) diminish at a rate that is proportional to the quantity of material present at that instant. If y denotes the quantity of radioactive material present at a time t then the rate at which the material decreases at that instant can be mathematically narrated as follows.

$$\frac{dy}{dt} \propto y$$

$$\therefore \quad \frac{dy}{dt} = -ky$$

The solution of the above equation will be $y = Y_0 e^{-kt}$ where Y_0 is the quantity of radioactive material present at $t = 0$.

Now let us find out the time τ at which the quantity of the radioactive material reduces to half of its initial value. This is popularly known as *half life*. In the above equation if we put $y = \frac{Y_0}{2}$ we get

$$\frac{Y_0}{2} = Y_0 e^{-k\tau}$$

$$\text{or, } e^{-k\tau} = \frac{1}{2}$$

Taking logarithm (base e) on both sides we get

$$\ln e^{-k\tau} = \ln 1 - \ln 2$$

$$\text{or, } -k\tau = 0 - \ln 2$$

$$\text{or, } \tau = \frac{\ln 2}{k}$$

2.21.2 Advance of Chemical Reaction

When a substance is dissolved in a comparatively large amount of solvent, in some cases a chemical reaction takes place following chemical law of mass action. As per this law the rate of reaction at any time t is proportional to the quantity of reacting substance still remaining unchanged.

Let $q(t)$ be the quantity of the substance which still remains unchanged at a time t. As per the law of mass action

$$\frac{dq(t)}{dt} \propto q(t)$$

$$\therefore \quad \frac{dq(t)}{dt} = -k.q(t),$$

where k is a constant depending on the reacting substance.

$$\therefore \quad q(t) = A.e^{-kt} \quad \text{where } A \text{ is a constant.}$$

At $t = 0$, $A = q(0) = Q_0$ (initial amount)

$$\therefore \quad q(t) = Q_0.e^{-kt}$$

2.21.3 Logistic growth

In 2.19 we assumed that a population $y(t)$ can grow exponentially to any extent without any bounds. In real world, however, there is a certain limit beyond which it can not grow since other factors (like lack of food) will play a restrictive role. In equation (2.40) we considered m as a constant; now we will replace it by $[a - by(t)]$ where a and b are constants. It is important to note that $[a - by(t)]$ decreases with increase in $y(t)$. Now we will consider the following.

$$\frac{dy(t)}{dt} = [a - b.y(t)]y(t) \qquad (2.49)$$

The solution of this differential equation is given below without a proof. Interested readers may find the proof in *Advanced Engineering Mathematics (2nd edition) — Michael D. Greenberg, Pearson Education.*

$$y(t) = \frac{a.y(0)}{[a - b.y(0)].e^{-at} + b.y(0)} \quad (2.50)$$

One should note that as t tends to ∞, e^{-at} tends to zero. As such, $y(t)$ can not have a value which is larger than $\frac{a}{b}$. This logistic model can be used to know the growth of various organisms, spread of diseases and rumours.

2.22 Matrix Exponential e^A

Let there be a single variable $x(t)$ and its first derivative $\dot{x}(t)$ is related to it in the following manner. The symbol $\dot{x}(t)$ is a concise representation of $\frac{dx(t)}{dt}$

$$\dot{x}(t) = Kx(t) \quad \text{where } K \text{ is a constant.}$$

We know that the solution to the above differential equation is

$$x(t) = B.e^{Kt} \quad \text{where } B \text{ is a constant.}$$

$$\text{At } t = 0, x(0) = B.e^{k.0} = B$$

$$\therefore \quad x(t) = x(0).e^{Kt} \quad (2.51)$$

Now if we have two variables $x_1(t)$ and $x_2(t)$ and the following two differential equations

$$\dot{x}_1(t) = a_{11}x_1(t) + a_{12}x_2(t)$$

$$\dot{x}_2(t) = a_{21}x_1(t) + a_{22}x_2(t)$$

The above two equations can be written in a more concise form as follows.

$$\dot{X}(t) = \dot{X}(t) = A.X(t) \quad (2.52)$$

where

$$\dot{X}(t) = \begin{bmatrix} \dot{x}_1(t) \\ \dot{x}_2(t) \end{bmatrix}; \quad A = \begin{bmatrix} a_{11} & a_{12} \\ a_{21} & a_{22} \end{bmatrix}; \quad X(t) = \begin{bmatrix} x_1(t) \\ x_2(t) \end{bmatrix}$$

It can be shown that the solution of the above equation is

$$X(t) = [I + A.t + A^2.\frac{t^2}{2!} + \cdots \infty].X(0) \quad (2.53)$$

Here

$$I = \begin{bmatrix} 1 & 0 \\ 0 & 1 \end{bmatrix}$$

All terms inside the bracket forming the infinite series are 2×2 matrices.

Matrix exponential e^{At} may now be defined as follows

$$e^{At} = I + A.t + A^2.\frac{t^2}{2!} + \cdots to \infty \qquad (2.54)$$

We should keep it in mind that here e^{At} represents a 2×2 matrix to which the R.H.S. infinite series converges.

We may now rewrite the equation for $X(t)$ in a more concise way.

$$X(t) = e^{At}.X(0) \qquad (2.55)$$

It is interesting to note that the solution for a single differential equation as described by the equation (2.51) and the solution for two differential equations as described by the equation (2.55) look alike.

Although we have not given any proof of equation (2.53) it can be justified by showing that it satisfies equation (2.52).

$$\begin{aligned}
\dot{X}(t) &= \frac{d[I + A.t + A^2.\frac{t^2}{2!} + \cdots \infty]}{dt}.X(0) \\
&= [0 + A + A^2.t + \frac{A^3}{2!}t^3 + \cdots to \infty].X(0) \\
&= A[I + A.t + A^2.\frac{t^2}{2!} + \cdots to \infty].X(0) \\
&= AX(t)
\end{aligned}$$

where O is a null matrix.

In equation (2.54) if we put $t = 1$ we get

$$e^A = I + A + A^2.\frac{1}{2!} + \cdots to \infty$$

This method of solving two coupled differential equations can also be extended to n number of coupled differential equations. In such a case e^A represents a $n \times n$ matrix. The infinite series of R.H.S very often converges very slowly; as such, special techniques are used to compute e^A.

2.23 Chronology of development of concepts related to e

- Simon Stevinus, Belgian engineer and financier, published compound interest tables in a book entitled *Tafelen van Interest, Midtsgaders De Constructie der seiner* [Tables of Interest together with their construction], Antwerp, 1582. In this book he mentioned that French Mathematician Jean Trenchant developed these Tables in 1558.

- Nicolaus Mercator gave the series expression of $\ln(1 + x)$ [*ln* stands for logarithm to base e] in his book *Logarithmotechnia* published in 1668. But he did not mention about this number e.

- Jacob Bernoulli defined e as the $\lim_{n \to \infty}(1 + \frac{1}{n})^n$ in 1689 while dealing with the problem of compound interest.

- In 1690 Leibnitz used a notation b in his letter to Hygens to denote the number e. Leibnitz and Johann Bernoulli found series expansions for e^x

- Johann Bernoulli included the calculus of exponential functions in his work *Principia calculi exponentialium seu percurrentium* published in 1697.

- In 1728, Euler introduced e for the base of natural logarithms in an unpublished manuscript (*Opera Posthuma*, II, 800-804).

- In 1731 Euler used the symbol e (which is currently used) in a letter to Goldbach.

  item Euler (i) expressed e as continued fraction (ii) in 1737 proved that e and e^2 are irrational (iii) if x is rational e^x will be irrational.

- Liouville in 1844 proved that e does not satisfy any quadratic equation with integral coefficients.

- Hermite subsequently settled the issue, proving e to be transcendental in 1873.

- Boorman calculated e to 346 decimal places in 1884

- e and e^x are extensively used in physics and engineering.

Bibliography

1. http://www-history.mcs.st-and.ac.uk/HistTopics/e.html

2. Zeldovich, Ya. B. and Yaglom,I. M., 1987 *Higher Math for Beginners.* Mir Publishers, Moscow

3. Cornelius Lanczos, 1968 *Numbers Without End.*   Oliver & Boyd, Edinburgh

4. Eli Maor, 1994 *e- the story of a number.* Princeton University Press, Princeton, New Jersey, USA

5. Lancelot Hogben, 1968 *Mathematics for the million, 4th edition.* W. W. Norton and Company, New York

6. Courant, R. 1937 *Differential and Integral Calculus, Vol. I, 2nd edition.* Blackie and Son Limited, Glasgow

7. Otto Toeplitz, 2007 *The Calculus, A Genetic Approach.* The University of Chicago Press, Chicago — London, Published in Association with the Mathematical Association of America

8. Bell, A. H., 1932 *The exponential and hyperbolic functions and their applications.* Sir Isaac Pitman & Sons Ltd., London

9. Tobias Dantzig *Number: The Language of Science.*

10. Paul J. Nahin, 1998 *The story of* $\sqrt{-1}$. Universities Press

11. Jyoti Prasanna Roy,1993 *Ganiter Bornomoy Bhumika* (In Bengali). Naba Granthana, Kolkata-6

12. Piskunov N., 1969 *Differential and Integral Calculus.* Mir Publishers, Moscow

13. Mallik, Asok Kumar, 2018 *The Story of Numbers* IISc Press

Logarithm

"It may come as a surprise to many that often times mathematical concepts don't end up like they started! For those of you who think mathematics is timeless, fixed, and full of unchanging truths, such a proposition may seem unbelievable. But there are many instances in the history of mathematics of the development of a mathematical concept way beyond the purposes and potentialities that its original inventors intended. An example that will be familiar to you all is the logarithm."

Kathleen M. Clark (The Florida State University) and Clemency Montelle (University of Canterbury) in *The Early History of a familiar Function*

"Seeing there is nothing that is so troublesome to mathematical practice, nor doth more molest and hinder calculators, than the multiplications, divisions, square and cubical extractions of great numbers, which besides the tedious expense of time are for the most part subject to many slippery errors, I began therefore to consider in my mind by what certain and ready art I might remove those hindrances. ⋯ Cast away from the work itself even the very numbers themselves that are to be multiplied, divided, and resolved into roots, and putteth other numbers in their place which perform much as they can do, only by addition and subtraction, division by two or division by three."

John Napier, Canon of Logarithms, 1614 as quoted in *When Slide Rules Ruled* by Cliff Stoll, *Scientific American Magazine, May 2006, pgs. 81*

3.1 Introduction

Logarithm was used mainly as a labour saving technique before the advent of computers while doing computational works related to a wide variety of scientific and engineering problems. Main impetus in this regard came from astronomy where it was frequently necessary to multiply and divide large numbers. However, logarithm can be perceived from many other angles. It can be viewed as the area under the rectangular hyperbola $y = \frac{1}{x}$ in geometry. It can be used as the inverse of exponential function e^x or a^x. As such we may treat it as the inverse of continuous compounding problem when we are interested to know in how many years Rs. 1/- will have a matured value e^x or a^x. In analysis we find that it is the limit of the product of two factors which are functions of n when n tends to infinity. It can also be expressed as an infinite series. It is one of the core functions in mathematics extended to negative and complex numbers. It plays vital roles in many branches of mathematics. Mathematical expressions for inductance and capacitance of a transmission line contain logarithmic terms. Logarithm forms the basis of Richter scale and measure of pH. It has wide applications in many other fields as well.

3.2 Logarithm as artificial numbers facilitating computation

"Logarithms are a set of artificial numbers invented and formed into tables for the purpose of facilitating arithmetical computations. They are adapted to the natural numbers in such a manner that by their aid Addition supplies the place of Multiplication, Subtraction to that of Division, Multiplication that of Involution, and Division that of Evolution or the Extraction of Roots".

Excerpt from *A Manual of Logarithms and Practical Mathematics for the use of students, Engineers, Navigators and Surveyors* — by James Trotter of Edinburgh Published by Oliver & Boyd, Tweeddale Court and Simpkin, Marshall, & Co. London in 1841.

In eleventh century Ibon Jonuis, an Arab mathematician proposed a method of multiplication which can save computational labour significantly. The method is known as Prosthaphaeresis. The Greek word prosthesis means addition and aphaeresis means subtraction. In sixteenth century two Danish mathematicians, Wittich and Clavius suggested similar methods using trigonometric tables to reduce computation during multiplication. The method is illustrated below through a numerical example.

Suppose we want to find the product of 0.45 and 0.95.

$$
\begin{aligned}
0.45 \times 0.95 &= \sin 26.74^\circ \times \cos 18.19^\circ \\
&= \frac{\sin(26.74^\circ + 18.19^\circ) + \sin(26.74^\circ - 18.19^\circ)}{2} \\
&= \frac{\sin 44.93^\circ + \sin 8.55^\circ}{2} \\
&= \frac{0.706 + 0.149}{2} = 0.4275
\end{aligned}
$$

So if we use trigonometric tables we may perform multiplication by adding (or sometimes subtracting) numbers and dividing the result by 2. Stifel, a fifteenth century German mathematician, noticed some interesting relationships exist between the terms of a G.P. series e.g., $1, r, r^2, r^3, \cdots$ and the exponent (power) of each term forming an A.P. series: $0, 1, 2, 3, \cdots$ Let us take a concrete example for $r = 2$ (refer Table 3.1).

Table 3.1: Numbers in G.P. and their exponents in A.P.

Numbers in G.P.	Numbers Expressed as 2^i	Exponent i in A.P.
1	2^0	0
2	2^1	1
4	2^2	2
8	2^3	3
16	2^4	4
32	2^5	5
64	2^6	6
128	2^7	7
256	2^8	8
512	2^9	9
1024	2^{10}	10
2048	2^{11}	11
4096	2^{12}	12

Stifel's observations are as follows.

1. Multiplication of two terms of column 1 is associated with addition of corresponding terms in column 3. Example: $8 \times 32 = 2^3 \times 2^5 = 2^{3+5} = 2^8 = 256$

2. Division of two terms of column 1 is associated with subtraction of corresponding terms in column 3. Example: $128 \div 32 = 2^7 \div 2^5 = 2^{(7-5)} = 2^2 = 4$

3. Raising the power of a term of column 1 is associated with multiplication of the corresponding term in column 3. Example: $(4)^3 = (2^2)^3 = 2^{2 \times 3} = 2^6 = 64$

4. Lowering the power of a term of column 1 is associated with division of the corresponding term in column 3. Example: $\sqrt[4]{256} = (256)^{\frac{1}{4}} = (2^8)^{\frac{1}{4}} = 2^{\frac{8}{4}} = 2^2 = 4$

Stifel, however, did not proceed further.

It is also possible to find out fractional powers of 2 which approximately produce numbers like 1.1, 1.2, 1.3 $\cdots$

$2^{0.138} \approx 1.1$; $2^{0.264} \approx 1.2$; $2^{0.379} \approx 1.3$; $2^{0.486} \approx 1.4$; $2^{0.585} \approx 1.5$; $2^{0.679} \approx 1.6$; $2^{0.766} \approx 1.7$; $2^{0.849} \approx 1.8$; $2^{0.926} \approx 1.9$; $2^{1.0} = 2.0$

We may proceed further indefinitely.

Based on this basic concept Napier, Burgi and Briggs, in the earlier part of 17th century developed Tables which made a veritable revolution in the field of computation involving large numbers. The actual methods used by them, however, demand a long discussion which we are not going to undertake now. They put hard labour to develop the Tables. Once the Tables were developed people could do a large class of computation with greater ease. How to use Log tables is taught in High School mathematics. As such, we are not going to discuss this now. The advent of calculator and computer has made log tables obsolete in our real life. But we must not forget that these Log Tables gave a great service to the scientific and engineering community for more than three hundred years. Using infinite series which was derived later it was possible to calculate logarithm of a number with less effort.

3.3 Logarithmic Function as an integral

Let us now consider the equation $2^x = y$ for which we may plot a curve for y for different values of x. We may, however, consider a more

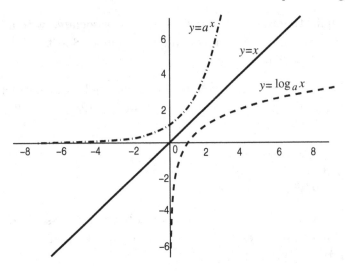

Figure 3.1: Curves for $y = a^x$ and $y = \log_a x$

general equation $a^x = y$ where a may take any positive real value. A typical curve for y for different values of x is shown in Figure 3.1. So y is a function of x which can be symbolically represented by $y = f_a(x)$. We may again think of an inverse functional relationship which expresses x as a function of y using the equation $x = f_a^{inv}(y)$. In this case y is considered as the independent variable and x as the dependent variable. Since x is usually considered the independent variable and y as the dependent variable we may interchange x and y to express the inverse function in the conventional way to obtain $y = f_a^{inv}(x)$. To remain in conformity with the widely used practice we may now replace f_a^{inv} by $\log_a$ which finally gives us $y = \log_a(x)$. Figure 3.1 shows the variation of the function a^x as well as its inverse function $\log_a x$ with the change in independent variable x. When e is used as the base (it means $a = e$) we write $y = \ln(x)$.

In nature many growths and decays follow a pattern which can be mathematically described by exponential equation of the form $y = e^{mx}$. As such, e is widely used as a base for logarithmic function. The integer number 10 is another commonly used base for logarithmic functions.

When $y = \ln x$, $x = e^y$, $\frac{dx}{dy} = e^y = x$

Using Chain rule, we can write,

$$1 = \frac{dy}{dy} = \frac{dy}{dx}.\frac{dx}{dy}$$

$$\therefore \quad \frac{dy}{dx} = \frac{1}{\frac{dx}{dy}} = \frac{1}{x}$$

$$\therefore \quad \ln x = y = \int \frac{dy}{dx}.dx = \int \frac{1}{x}.dx$$

So we find that logarithm can be perceived as the area under the curve $\frac{1}{x}$.

Alternate proof:

When $y = \ln x$, $x = e^y = e^{\ln x}$

Differentiating both sides,

$$\frac{dx}{dx} = 1 = \frac{de^{\ln x}}{dx} = \frac{de^{\ln x}}{d(\ln x)}.\frac{d(\ln x)}{dx} = e^{\ln x}.\frac{d(\ln x)}{dx} = x.\frac{d(\ln x)}{dx}$$

$$\therefore \quad \frac{d \ln x}{dx} = \frac{1}{x}$$

$$\therefore \quad \ln x = \int \frac{1}{x}.dx$$

Logarithm with any base a:

Let $f(x) = a^x$ and $a = e^m$. Now

$$
\begin{aligned}
f'(x) &= \lim_{h \to 0} \frac{f(x+h) - f(x)}{h} \\
&= \lim_{h \to 0} \frac{a^{x+h} - a^x}{h} \\
&= a^x \cdot \lim_{h \to 0} \frac{a^h - a^0}{h} = a^x f'(0)
\end{aligned}
$$

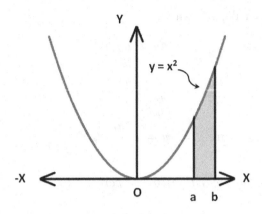

Figure 3.2: Area under the parabola $y = x^2$

Let $y = \log_a x$

$$\therefore \quad a^y = e^{my} = x$$

$$\text{or, } my = \ln x$$

$$\text{or, } y = \frac{1}{m}\ln x$$

$$\text{or, } \log_a x = \frac{1}{m}\ln x = \frac{1}{\ln a}\int \frac{1}{x}.dx = \int \frac{k}{x}.dx$$

where $k = \frac{1}{\ln a}$

3.4   A story of the historical development of logarithm as an area

In ancient Greece, Archimedes devised a method using which it is possible to find the area under a parabola $y = x^2$ (refer Figure 3.2). Using present day notation we may write it as follows.

$$A_2 = \int_a^b x^2.dx = \frac{b^3 - a^3}{3}$$

Later in Europe, Cavalieri extended this search further around 1630 for area under the curves $y = x^3$, $y = x^4$,.....upto $y = x^9$. He, however, got stuck for the case $y = x^{10}$.

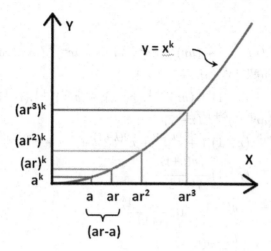

Figure 3.3: Area under the curve $y = x^k$

Around 1650, Fermat derived a general result for $y = x^k$.

$$A_k = \int_a^b x^k \cdot dx = \frac{b^{k+1} - a^{k+1}}{k+1} \tag{3.1}$$

Let us now derive this equation. We first divide the interval $(b-a)$ into n numbers of subdivisions (refer Figure 3.3) such that these subdivisions are in geometric progression. We choose a common ratio r. So the subdivisions will be as follows.

$$a.r^0(= a),\, ar,\, ar^2,\, ar^3,\, ...,\, ar^{n-1},\, ar^n(= b)$$

Since $ar^n = b$, $\quad \therefore\ r = \sqrt[n]{\frac{b}{a}}$

To derive the R.H.S. of equation (3.1) we have to find out the following sum for a very large value of n.

$$
\begin{aligned}
S_k &= a^k.(ar - a) + (ar)^k.(ar^2 - ar) + (ar^2)^k.(ar^3 - ar^2) + \cdots \\
&\quad + (ar^{n-1})^k.(ar^n - ar^{n-1}) \\
&= a^{k+1}.(r-1) + (ar)^{k+1}.(r-1) + (ar^2)^{k+1}.(r-1) + \cdots \\
&\quad + (ar^{n-1})^{k+1}.(r-1) \\
&= a^{k+1}.(r-1)[1 + r^{k+1} + r^{2(k+1)} + \cdots + r^{(n-1)(k+1)}] \\
&= a^{k+1}.(r-1).\frac{r^{n(k+1)} - 1}{r^{(k+1)} - 1} \\
&= a^{k+1}.[r^{n(k+1)} - 1].\frac{(r-1)}{r^{(k+1)} - 1} \\
&= a^{k+1}.[(\frac{b}{a})^{\frac{1}{n}n(k+1)} - 1].\frac{1}{\frac{r^{(k+1)}-1}{(r-1)}} \\
&= (b^{k+1} - a^{k+1}).\frac{1}{r^k + r^{k-1} + r^{k-2} + \cdots + r + 1} \quad (3.2)
\end{aligned}
$$

If we subdivide the interval $(b - a)$ into infinite number of subdivisions then $r = \lim_{n \to \infty}(\frac{b}{a})^{\frac{1}{n}} = (\frac{b}{a})^0 = 1$

$$
\begin{aligned}
\therefore \quad S_k &= (b^{k+1} - a^{k+1}).\frac{1}{1 + 1 + 1 + \cdots + 1 + 1} \\
&= (b^{k+1} - a^{k+1}).\frac{1}{k+1} = \frac{b^{k+1} - a^{k+1}}{k+1} \quad (3.3)
\end{aligned}
$$

It is important to note that the above equation is valid for any value of k except when $k = -1$ *which yields* $S_{-1} = \frac{0}{0}$. *So people could not find* $S_{-1} = \int_a^b \frac{1}{x}.dx$

However, Jesuit Father Gregorius a Santo Vincenito made an important discovery in 1647 which can be stated in modern form as follows.

$$
\int_a^b \frac{1}{x}.dx = \int_{ma}^{mb} \frac{1}{x}.dx \quad (3.4)
$$

In other words, if the limits of integration are multiplied by the same number m the result remains the same.

Justification (non-rigorous) of the above statement:

Let us consider a particular case where $a = 1$, $b = 2$ and $m = 2$

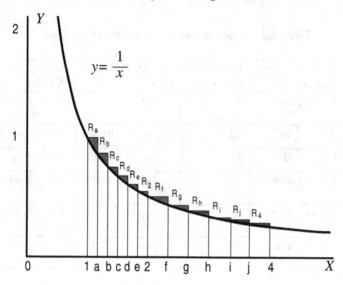

Figure 3.4: Area under the curve $y = \frac{1}{x}$

We will now have to justify the following

$$\int_1^2 \frac{1}{x} \, . \, dx = \int_2^4 \frac{1}{x} \, . \, dx$$

Let us divide the range $(1-2)$ (refer Figure 3.4) into six equal parts viz., $(1-a)$, $(a-b)$, $(b-c)$, $(c-d)$, $(d-e)$, and $(e-2)$. Each part will have a magnitude of $\frac{1}{6}$ unit. We also divide the range $(2-4)$ into six equal parts viz., $(2-f)$, $(f-g)$, $(g-h)$, $(h-i)$, $(i-j)$ and $(j-4)$. Each part will have a magnitude of $\frac{1}{3}$ unit (refer Figure 3.4).

We will now find out the areas of rectangles R_a, R_b, R_c, R_d, R_e and R_2 in Table 3.2.

We will now find out the areas of rectangles R_f, R_g, R_h, R_i, R_j and R_4 in Table 3.3.

Comparing Table 3.2 and Table 3.3 we find that

Area of Rectangle R_a = Area of Rectangle R_f
Area of Rectangle R_b = Area of Rectangle R_g
Area of Rectangle R_c = Area of Rectangle R_h
Area of Rectangle R_d = Area of Rectangle R_i
Area of Rectangle R_e = Area of Rectangle R_j
Area of Rectangle R_2 = Area of Rectangle R_4

Table 3.2: Approximate computation of $\int_1^2 \frac{1}{x}.dx$

Sl No.	Width(along X-axis)	Rectangle Name	Length (along Y-axis)	Area (Length × Width)
1	1-a $= \frac{1}{6}$unit	R_a	1 unit	$A_a = \frac{1}{6}$ sq. unit
2	a-b $= \frac{1}{6}$unit	R_b	$\frac{1}{(1+\frac{1}{6})} = \frac{6}{7}$unit	$A_b = \frac{1}{7}$ sq. unit
3	b-c $= \frac{1}{6}$unit	R_c	$\frac{1}{(1+\frac{2}{6})} = \frac{6}{8}$unit	$A_c = \frac{1}{8}$ sq. unit
4	c-d $= \frac{1}{6}$unit	R_d	$\frac{1}{(1+\frac{3}{6})} = \frac{6}{9}$unit	$A_d = \frac{1}{9}$ sq. unit
5	d-e $= \frac{1}{6}$unit	R_e	$\frac{1}{(1+\frac{4}{6})} = \frac{6}{10}$unit	$A_e = \frac{1}{10}$ sq. unit
6	e-2 $= \frac{1}{6}$unit	R_2	$\frac{1}{(1+\frac{5}{6})} = \frac{6}{11}$unit	$A_2 = \frac{1}{11}$ sq. unit

Table 3.3: Approximate computation of $\int_2^4 \frac{1}{x}.dx$

Sl No.	Width(along X-axis)	Rectangle Name	Length (along Y-axis)	Area (Length × Width)
1	2-f $= \frac{1}{3}$unit	R_f	$\frac{1}{2}$ unit	$A_f = \frac{1}{6}$ sq. unit
2	f-g $= \frac{1}{3}$unit	R_g	$\frac{1}{(2+\frac{1}{3})} = \frac{3}{7}$unit	$A_g = \frac{1}{7}$ sq. unit
3	g-h $= \frac{1}{3}$unit	R_h	$\frac{1}{(2+\frac{2}{3})} = \frac{3}{8}$unit	$A_h = \frac{1}{8}$ sq. unit
4	h-i $= \frac{1}{3}$unit	R_i	$\frac{1}{(2+\frac{3}{3})} = \frac{3}{9}$unit	$A_i = \frac{1}{9}$ sq. unit
5	i-j $= \frac{1}{3}$unit	R_j	$\frac{1}{(2+\frac{4}{3})} = \frac{3}{10}$unit	$A_j = \frac{1}{10}$ sq. unit
6	j-4 $= \frac{1}{3}$unit	R_4	$\frac{1}{(2+\frac{5}{3})} = \frac{3}{11}$unit	$A_4 = \frac{1}{11}$ sq. unit

$$\therefore \ A_a + A_b + A_c + A_d + A_e + A_2 = A_f + A_g + A_h + A_i + A_j + A_4$$

The sum of the LHS areas $= A_a + A_b + A_c + A_d + A_e + A_2$

$$= \int_1^2 \frac{1}{x}.dx + \epsilon_1$$

The sum of the RHS areas $= A_f + A_g + A_h + A_i + A_j + A_4$

$$= \int_2^4 \frac{1}{x} . dx + \epsilon_2$$

Here ϵ_1 and ϵ_2 are the errors introduced in the sum of areas due to the existence of portions of rectangles lying above the curve $y = \frac{1}{x}$. Instead of dividing the range (1–2) and (2–4) into six parts if we divide it into very large number of parts top of the resultant rectangles will almost coincide with the curve $y = \frac{1}{x}$ and no portion of any rectangle will lie above the curve. As such $\epsilon_1 \approx 0$ and $\epsilon_2 \approx 0$.

$$\therefore \quad \int_1^2 \frac{1}{x} . dx = \int_2^4 \frac{1}{x} . dx$$

However, we can prove it easily using the technique of Calculus. Let us introduce a new variable $z = mx$

$$\text{or,} \quad x = \frac{z}{m} \text{ and } dx = \frac{dz}{m}$$

When $x = a$, $z = ma$
When $x = b$, $z = mb$

$$\therefore \quad \int_a^b \frac{1}{x} . dx = \int_{ma}^{mb} \frac{m}{z} . \frac{dz}{m} = \int_{ma}^{mb} \frac{1}{z} . dz$$

It is important to point out here that the variable of integration is a dummy variable and can be denoted by any symbol x, z or anything else.

3.5 Reverse Problem

We will now consider the reverse problem for a specific case when $a = 1$ and area is A for an upper limit b. We will then see whether there is any mathematical relationship between A and b.

Area of the first curvilinear trapezium $= (r - 1).1 = (r - 1)$ (refer Figure 3.6)

$\therefore$ n trapeziums have an area $= n.(r - 1)$

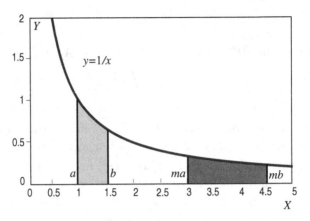

Figure 3.5: Equal areas

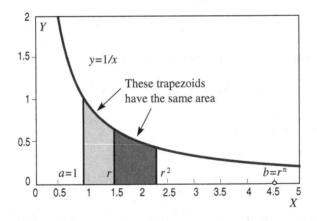

Figure 3.6: Figure related to section 3.5

But we know,

$$\int_1^b \frac{1}{x}.dx = A$$

$\therefore \ n.(r-1) = A$

or, $r - 1 = \frac{A}{n}$

or, $b^{\frac{1}{n}} - 1 = \frac{A}{n}$

or, $b = (1 + \frac{A}{n})^n$

When $n \to \infty$, R.H.S. of the above equation $\to e^A$

$$\therefore \quad b = e^A \tag{3.5}$$

So we know that the upper limit $b = f(A) = e^A$ where $f(A)$ denotes function of A.

We can also introduce a symbol f^{inv} to express area A in terms of upper limit of integration $b = e^A$.

$$\therefore \quad A = f^{inv}(b) \tag{3.6}$$

It is customary to use $\ln b$ *instead of* $f^{inv}(b)$.

$$\therefore \quad A = \ln b \tag{3.7}$$

Table 3.4: Area A for different values of b

Abscissa $x = b = e^A$	Area $A = \int_1^b \frac{1}{x} dx$
$x_1 = e^1$	1
$x_1 = e^2$	2
$x_1 = e^3$	3
$x_1 = e^4$	4
.	.
.	.

From Table 3.4 it is clear that when elements x_i are in Geometric Progression the corresponding elements A_i are in Arithmetic Progression.

3.6 Some useful properties of logarithmic functions

$$1. \quad \int_1^1 \frac{1}{x} . dx = 0 = \ln 1 \tag{3.8}$$

$$2. \quad \int_1^b \frac{1}{x} . dx = \ln b \tag{3.9}$$

3. $\ln(uv) = \ln u + \ln v$ (3.10)

$[R.H.S. = \int_1^u \frac{1}{x}.dx + \int_1^v \frac{1}{x}.dx$

$\qquad = \int_1^u \frac{1}{x}.dx + \int_u^{uv} \frac{1}{x}.dx$

$\qquad = \int_1^{uv} \frac{1}{x}.dx = \ln(uv) = L.H.S.]$

4. $\ln\left(\dfrac{u}{v}\right) = \ln u - \ln v$ (3.11)

$[\ln 1 = \ln(v.\frac{1}{v}) = \ln v + \ln \frac{1}{v} = 0$

$\therefore \ \ln \frac{1}{v} = -\ln v$

$\therefore \ \ln \frac{u}{v} = \ln(u.\frac{1}{v}) = \ln u + \ln \frac{1}{v} = \ln u - \ln v]$

5. $\ln u^n = n \ln u$ (3.12)

when n is an integer

[By generalising the result of 3, we may infer that the logarithm of product of two, three, four $\cdots$ numbers will be the sum of the logarithms of those numbers.

For an integer n,

$\ln u^n = \ln(u.u.u \cdots u) = \ln u + \ln u + \ln u + \cdots + \ln u = n \ln u]$
$\qquad \leftarrow n \ numbers \rightarrow$

6. $\ln \sqrt[n]{u} = \dfrac{1}{n}.\ln u$ (3.13)

$[\text{ Let } v = \sqrt[n]{u} \ \ \therefore \ v^n = u$

$\therefore \ \ln u = \ln v^n = n.\ln v = n.\ln \sqrt[n]{u}$

$\therefore \ \ln \sqrt[n]{u} = \frac{1}{n}.\ln u \ ]$

7. $\ln u^n = n.\ln u$ (3.14)

when n is a rational number.

$[\text{Let } n = \frac{p}{q} \ \ \therefore \ \ln u^{\frac{p}{q}} = \ln \sqrt[q]{u^p} = \frac{1}{q}\ln u^p = \frac{1}{q}p.\ln u = \frac{p}{q}.\ln u]$

8. $\ln u^n = n.\ln u$ (3.15)

when n is a negative number.

$[\text{Let } k = -n \ \ \therefore \ u^n = \frac{1}{u^{-n}} = \frac{1}{u^k}$

$\therefore \ \ln u^n = \ln 1 - \ln u^k = 0 - k \ln u = n \ln u]$

9. $\qquad \ln u^n = n.\ln u$ $\qquad\qquad$ (3.16)

when n is a zero.

$[\ \ln u^0 = \ln 1 = 0 = 0.\ln u\ ]$

From equation (3.12), equation (3.14), equation (3.15) and equation (3.16) it is clear that for any rational n (+ve, -ve or zero) $\ln u^n = n.\ln u$. *This relationship is also valid when* n *is irrational. We will accept this statement without proof.*

3.7 Expressing logarithm as a series

We know that

$$\int_1^b \frac{1}{x} dx = \ln b$$

$\therefore$ When $b = 1 + p$,

$$\int_1^{1+p} \frac{1}{x} dx = \ln(1 + p)$$

In Figure 3.7, let $x = z + 1$ (physically it means the vertical axis Y is shifted right by $OO' = 1$ unit so that at $z = 0, x = 1$).

$\therefore$ $dx = dz$; $\qquad$ when $x = 1, z = 0$; and when $x = 1 + p, z = p$,

$$\therefore \int_1^{1+p} \frac{1}{x} dx = \int_0^p \frac{1}{z+1} dz = \ln(1+p)$$

We know that the sum of the G.P series up to $2n$ number of terms present at the L.H.S of the following equation with 1 as its 1st term, $(-z)$ as its common ratio and $(-z^{2n-1})$ as its last term can be worked out as follows:

$$1 - z + z^2 - z^3 + z^4 - z^5 + \dots\dots - z^{2n-1} = \frac{1-z^{2n}}{1+z} = \frac{1}{z+1} - \frac{z^{2n}}{z+1}$$

$$\therefore \quad \frac{1}{z+1} = 1 - z + z^2 - z^3 + z^4 - z^5 + \cdots - z^{2n-1} + \frac{z^{2n}}{z+1}$$

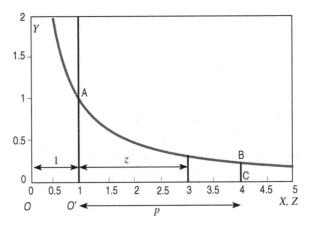

Figure 3.7: Figure related to section 3.7

$$\therefore \ \ln(1+p) \ = \ \int_0^p \frac{1}{z+1} \, dz$$

$$= \ \int_0^p [1 - z + z^2 - z^3 + z^4 - z^5 + \cdots - z^{2n-1} + \frac{z^{2n}}{z+1}] \, dz$$

$$= \ \int_0^p 1.dz - \int_0^p z.dz + \int_0^p z^2.dz \cdots$$

$$- \int_0^p z^{2n-1}.dz + \int_0^p \frac{z^{2n}}{z+1}.dz$$

$$= \ (p - \frac{p^2}{2} + \frac{p^3}{3} - \frac{p^4}{4} + \cdots - \frac{p^{2n}}{2n}) + \int_0^p \frac{z^{2n}}{z+1}.dz \quad (3.17)$$

Now suppose the value of p is known. We can choose a large value of n to calculate $\ln(1+p)$ using only the polynomial present in side the parenthesis of the R.H.S. of equation (3.17) without considering the integral outside the parenthesis. For a large value of n the result may be fairly accurate. However, there will always be an error which can be mathematically expressed by the integral $\int_0^p \frac{z^{2n}}{z+1}.dz$.

$$\therefore \ \ln(1+p) \approx p - \frac{p^2}{2} + \frac{p^3}{3} - \frac{p^4}{4} + \cdots - \frac{p^{2n}}{2n} \quad (3.18)$$

To know the error associated with this approximate equation we

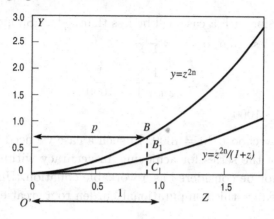

Figure 3.8: Simplifying error calculation

need to find out the deleted integral $\int_0^p \frac{z^{2n}}{z+1}.dz$. Since it is a difficult task we will now simplify the problem of error calculation.

Let us choose a p which satisfies the condition $0 < p \le 1$. For such values of p in the range $z = o$ to $z = p$ the following mathematical relation holds good.

$$0 < \frac{z^{2n}}{z+1} < z^{2n}$$

To visualize the above expression we may refer to the Figure 3.8 and note that the curve $y = \frac{z^{2n}}{z+1}$ lies below the curve $y = z^{2n}$ in the specified zone of p.

$$\therefore \quad \text{The area } O'CB_1 < \text{The area } O'CB$$

$$\int_0^p \frac{z^{2n}}{z+1}.dz < \int_0^p z^{2n}.dz = \frac{p^{2n+1}}{2n+1} \qquad (3.19)$$

So the actual error will be somewhat less than $\frac{p^{2n+1}}{2n+1}$ in the region $0 < p \le 1$. However, we can make it smaller and smaller by choosing larger and larger values of n. Moreover, as $n \to \infty$ the error tends to zero. As such, the infinite series $p - \frac{p^2}{2} + \frac{p^3}{3} - \frac{p^4}{4} + \cdots - \frac{p^{2n}}{2n}$ for $\ln(1 + p)$ is a converging series.

Let us now consider a concrete example. For $p = 1$ we get,

$$\ln 2 \approx 1 - \frac{1}{2} + \frac{1}{3} - \frac{1}{4} + \frac{1}{5} - \frac{1}{6} + \cdots + \frac{1}{2n-1} - \frac{1}{2n}$$

The error for such a case will be less than $\frac{1}{2n+1}$. Suppose we want to keep error less than $0.001 = \frac{1}{1000}$.

$$\therefore \quad \frac{1}{2n+1} < \frac{1}{1000}$$

or, $(2n+1) > 1000$

So we may choose $2n = 1000$. In such a case we need to compute 1000 terms of the series for achieving an accuracy within 0.001. As such, it cannot be considered as a workable equation. However, it is possible to reduce this computational burden to a great extent.

3.7.1 Alternate Method of Derivation of Logarithmic Series

3.7.1.1 a^x series:

$$e^y = 1 + y + \frac{y^2}{2!} + \frac{y^3}{3!} + \cdots$$

Let $y = cx$

$$\therefore \quad e^{cx} = 1 + cx + \frac{c^2 x^2}{2!} + \frac{c^3 x^3}{3!} + \cdots$$

Let $e^c = a$ $\therefore$ $c = \ln a$

$$\therefore \quad a^x = 1 + x.\ln a + x^2.\frac{(\ln a)^2}{2!} + x^3.\frac{(\ln a)^3}{3!} + \cdots \qquad (3.20)$$

3.7.1.2 $\ln(1+x)$ series:

$$a^z = 1 + z.\ln a + z^2.\frac{(\ln a)^2}{2!} + z^3.\frac{(\ln a)^3}{3!} + \cdots$$

Let $a = (1+x)$

$$\therefore \quad (1+x)^z = 1 + z.\ln(1+x) + z^2.\frac{[\ln(1+x)]^2}{2!} + z^3.\frac{[\ln(1+x)]^3}{3!} + \cdots$$
$$(3.21)$$

Using Binomial Theorem for $x < 1$ we get

$$\therefore \quad (1+x)^z = 1 + z.x + \frac{z(z-1)}{2!}.x^2 + \frac{z(z-1)(z-2)}{3!}.x^3 + \cdots \quad (3.22)$$

Comparing the coefficients of z of equation (3.21) and equation (3.22) we can write

$$
\begin{aligned}
\ln(1+x) &= x - \frac{x^2}{2} + \frac{(-1)(-2)}{1.2.3}x^3 + \frac{(-1)(-2)(-3)}{1.2.3.4}x^4 + \cdots \\
&= x - \frac{x^2}{2} + \frac{x^3}{3} - \frac{x^4}{4} + \cdots \qquad (3.23)
\end{aligned}
$$

3.7.2 Searching for a method with a higher rate of convergence

We will once again consider the integral

$$
\int_0^p \frac{z^{2n}}{z+1} . dz
$$

but now we will consider a negative p lying in between 0 and -1 i.e., $-1 < p < 0$.

We know that

$$
\int_0^p \frac{z^{2n}}{z+1} . dz = - \int_p^0 \frac{z^{2n}}{z+1} . dz
$$

In Figure 3.9 the hatched area $BCO'K$ shows the integral $\int_p^0 \frac{z^{2n}}{z+1} . dz$. It is evident that this area is a positive quantity. The integral $\int_0^p \frac{z^{2n}}{z+1} . dz$ will give a result having the same magnitude with a negative sign.

As such,

$$
\int_p^0 \frac{z^{2n}}{z+1} . dz = \left| \int_0^p \frac{z^{2n}}{z+1} . dz \right|
$$

For $z > p$, $\quad 1 + z > 1 + p$

Again for $p > -1$, $\quad 1 + p > 0$

$\therefore$ when $z > p$ and $p > -1$, $\quad 1 + z > 1 + p > 0$

$\therefore \quad \frac{1}{1+z} < \frac{1}{1+p}$

Since $2n$ is an even number z^{2n} is positive for any value of z.

$\therefore \quad \frac{z^{2n}}{1+z} < \frac{z^{2n}}{1+p}$

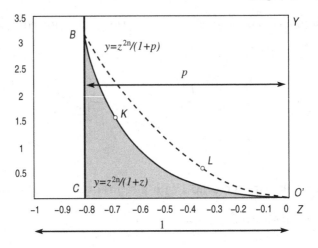

Figure 3.9: Simplifying error calculation as described in 3.7.2

It is also observed in Figure 3.9 that in the region $-1 < p < 0$ the curve representing the function $\frac{z^{2n}}{1+z}$ lies below the curve for $\frac{z^{2n}}{1+p}$.

$$\therefore \qquad \int_p^0 \frac{z^{2n}}{z+1}.dz < \int_p^0 \frac{z^{2n}}{p+1}.dz \qquad\qquad (3.24)$$

L.H.S. of the inequation (3.24) represents the error; however, it is difficult to evaluate this integral. On the contrary it is easier to evaluate the R.H.S integral. As such, we will use this integral while calculating the approximate value of $\ln(1 + p)$ using finite number of terms of the infinite series for the same. If we assess the error using the R.H.S integral it is assured that the actual error will be somewhat less than the calculated error. Let us now evaluate the R.H.S. of the inequality.

$$\int_p^0 \frac{z^{2n}}{p+1}.dz \;=\; \frac{1}{p+1}\int_p^0 z^{2n}.dz = \frac{1}{p+1}.\frac{0^{2n+1}-p^{2n+1}}{2n+1}$$

$$=\; \frac{-p^{2n+1}}{(2n+1)(p+1)} \qquad\qquad (3.25)$$

Since $-1 < p < 0$ and $2n+1$ is an odd number, p^{2n+1} is a negative quantity and $(p + 1)$ is positive. As such, the above integral is a positive quantity.

$$\therefore \quad \left| \int_0^p \frac{z^{2n}}{z+1} . dz \right| = \int_p^0 \frac{z^{2n}}{z+1} . dz < \frac{-p^{2n+1}}{(2n+1)(p+1)}$$

So if we neglect the term

$$\int_p^0 \frac{z^{2n}}{z+1} . dz$$

while calculating $\ln(1+p)$ there will be an error which is less than $\frac{-p^{2n+1}}{(2n+1)(p+1)}$ for p lying between -1 and 0. Moreover, as $n \to \infty$ the error tends to zero. As such, the infinite series $\ln(1+p) = p - \frac{p^2}{2} + \frac{p^3}{3} - \frac{p^4}{4} + \cdots$ is a converging series.

Let us choose $p = -\frac{1}{2}$ and $n = 4$

$$\therefore \quad \text{the error} \quad \frac{(\frac{1}{2})^9}{9.\frac{1}{2}} = \frac{1}{9 \times 256} = \frac{1}{2304} < 0.0005$$

So with this accuracy we can write,

$$\ln \left(\frac{1}{2} \right) \approx -\frac{1}{2} - \frac{1}{2^2.2} - \frac{1}{2^3.3} - \frac{1}{2^4.4} - \frac{1}{2^5.5} - \frac{1}{2^6.6} - \frac{1}{2^7.7} - \frac{1}{2^8.8}$$

However, while calculating the above terms we will have to round off the fractions which will again result in some rounding of errors. When we approximately represent each of the 8 summands into decimal fractions having 4 decimal places an error up to .00005 may occur.

So the overall error will not exceed 0.001 in this case.

$$\ln \left(\frac{1}{2} \right) \approx -0.5 - 0.125 - 0.0417 - 0.0156 - 0.0062$$
$$-0.0026 - 0.0011 - 0.0005$$
$$= -0.6927$$

As general procedure if we take $p = -\frac{k}{k+1}$ where k is an integer, we get $\ln(1 - \frac{k}{k+1}) = \ln 1 - \ln(k+1) = -\ln(k+1)$. As such, we can compute the approximate numerical value for $\ln(k+1)$ using a finite number of terms of equation (3.18). However, it is also important to know whether it is required to consider a large number of terms to keep the accuracy of result within a desired level.

Now let us take $k + 1 = 11$ or $k = 10$. Therefore, $p = -\frac{10}{11}$

Suppose we can tolerate an error which is less than 0.001.

From equation (3.25) we know that error in the approximate formula (3.18) for negative values of p is less than

$$\frac{(\frac{10}{11})^{2n+1}}{(2n+1)(1-\frac{10}{11})} = \frac{11}{2n+1} \cdot (\frac{10}{11})^{2n+1}$$

Table 3.5 shows the upper limit of error for different values of n.

Table 3.5: Upper limit of error for different values of n.

n	Upper limit of error
1	2.7548
2	1.3660
4	0.5183
8	0.1280
16	0.0143
28	0.0008

For $n = 28$, $2n = 56$. As such, we need to calculate the sum of 56 terms of the following equation to compute $\ln 11$.

$$\ln 11 = -\ln\left(\frac{1}{11}\right) = \frac{10}{11} + \frac{1}{2}\left(\frac{10}{11}\right)^2 + \frac{1}{3}\left(\frac{10}{11}\right)^3 + \cdots + \frac{1}{56}\left(\frac{10}{11}\right)^{56}$$

Since we still need to execute a large number of computational work it would be worthwhile to search for a method which requires even fewer computations.

3.7.3 Further improvement of the method described in section 3.7.2

Let $m = p$ when $0 < p \leq 1$, and $-m = p$ when $-1 < p < 0$.

We know that $\ln(1 + p) \approx p - \frac{p^2}{2} + \frac{p^3}{3} - \frac{p^4}{4} + \ldots\ldots - \frac{p^{2n}}{2n}$

In the above equation if we put $m = p$, we get

$$\ln(1 + m) \approx m - \frac{m^2}{2} + \frac{m^3}{3} - \frac{m^4}{4} + \cdots - \frac{m^{2n}}{2n} \qquad \text{when } 0 < m \leq 1$$
$$(3.26)$$

From equation (3.19) we know that error in this case will be less than $\frac{m^{2n+1}}{2n+1}$

If we put $-m = p$ in the above equation we get

$$\ln(1 - m) \approx -m - \frac{m^2}{2} - \frac{m^3}{3} - \frac{m^4}{4} - \cdots - \frac{m^{2n}}{2n} \qquad \text{when } 0 < m \leq 1$$
$$(3.27)$$

From equation (3.25) we know that error in this case will be less than $\frac{m^{2n+1}}{(2n+1)(1-m)}$

Subtracting equation (3.27) from equation (3.26) we get

$$\ln(1 + m) - \ln(1 - m) \approx 2\left(m + \frac{m^3}{3} + \frac{m^5}{5} + \frac{m^7}{7} + \cdots + \frac{m^{2n-1}}{2n - 1} \right)$$
$$\text{when } 0 < m \leq 1 \qquad (3.28)$$

Let $m = \frac{1}{2k+1}$ where k is any positive integer. As such, m will always be less than 1. If we replace m in equation (3.28) we get,

$$\ln(1 + \frac{1}{2k + 1}) - \ln(1 - \frac{1}{2k + 1})$$
$$\approx 2\left[\frac{1}{2k + 1} + \frac{1}{3.(2k + 1)^3} \right.$$
$$\left. + \frac{1}{5.(2k + 1)^5} + \cdots + \frac{1}{(2n - 1).(2k + 1)^{2n-1}} \right]$$
$$\text{when } 0 < m \leq 1 \qquad (3.29)$$

The L.H.S. of the above equation can be simplified as follows.

$$\ln\left(\frac{2k + 2}{2k + 1} \right) - \ln\left(\frac{2k}{2k + 1} \right) = \ln\left(\frac{k + 1}{k} \right) = \ln(k + 1) - \ln(k)$$

Therefore, equation (3.29) can be written as

$$\ln(k+1) - \ln(k) \approx 2\left[\frac{1}{2k+1} + \frac{1}{3.(2k+1)^3}\right.$$
$$\left. + \frac{1}{5.(2k+1)^5} + \cdots + \frac{1}{(2n-1).(2k+1)^{2n-1}}\right]$$

$$\therefore \quad \ln(k+1) \approx \ln(k) + 2\left[\frac{1}{2k+1} + \frac{1}{3.(2k+1)^3} + \frac{1}{5.(2k+1)^5}\right.$$
$$\left. + \cdots + \frac{1}{(2n-1).(2k+1)^{2n-1}}\right] \qquad (3.30)$$

As such, if we know the logarithm of any integer k we can find out the logarithm of the next integer $k+1$.

The absolute value of error in this case will not be more than the sum of errors of the formulas for $\ln(1 + \frac{1}{2k+1})$ and $\ln(1 - \frac{1}{2k+1})$ [refer equation (3.19) and equation (3.25); here $p = \frac{1}{2k+1}$] which is shown below.

$$\left[\frac{1}{2n+1} \cdot \frac{1}{(2k+1)^{2n+1}}\right] + \left[\frac{2k+1}{2k} \cdot \frac{1}{2n+1} \cdot \frac{1}{(2k+1)^{2n+1}}\right]$$

$$= \frac{1}{2n+1} \cdot \frac{1}{(2k+1)^{2n+1}} \cdot \frac{4k+1}{2k} < \frac{1}{2n+1} \cdot \frac{1}{(2k+1)^{2n+1}} \cdot \frac{4k+2}{2k}$$

Simplifying the R.H.S. of the above inequality we get

$$\frac{1}{k(2n+1).(2k+1)^{2n+1}}$$

So it is guaranteed that the absolute value of error will be less than

$$\frac{1}{k(2n+1).(2k+1)^{2n+1}}$$

3.7.3.1 Construction of Table of natural logarithm of integers

a. Since $\ln 1 = 0$ we can easily calculate $\ln 2$ using equation (3.30) and choosing $k = 1$ and $n = 5$ (it ensures an accuracy which is less than

$\frac{1}{2\times5+1}\cdot\frac{1}{(2\times1+1)^5} = \frac{1}{11}\cdot\frac{1}{3^{10}} < 0.000002)$ as follows.

$$\ln 2 \approx \ln 1 + 2\left[\frac{1}{3} + \frac{1}{3.3^3} + \frac{1}{5.3^5} + \frac{1}{7.3^7} + \frac{1}{9.3^9}\right] \approx 0.69315$$

b. If we take $k = 2$ and $n = 3$ we can calculate $\ln 3$

$\ln 3 \approx \ln 2 + 2\left[\frac{1}{5} + \frac{1}{3.5^3} + \frac{1}{5.5^5}\right] = \ln 2 + 0.4 + 0.005333 + 0.000128$
$= 0.69315 + 0.405461 = 1.09861$

Error in this case $< \frac{1}{2}\cdot\frac{1}{2\times3+1}\cdot\frac{1}{(2\times2+1)^6} = \frac{1}{2}\cdot\frac{1}{7}\cdot\frac{1}{5^6}$ which is less than 0.000005

c. To calculate the natural logarithm of prime numbers we need to use equation (3.30). Calculation of logarithm of composite numbers is, however, easier. We use the following equations for this purpose.

$$\ln(m \times n) = \ln m + \ln n$$

$$\ln m^k = k.\ln m$$

We can calculate $\ln 4, \ln 6, \ln 8, \ln 9, \cdots$ in this way. Once we know $\ln 4$ we can use equation 3.30 to calculate $\ln 5$ and if we know $\ln 6$ we can calculate $\ln 7$ using the same equation.

3.8 Logarithmic curves

3.8.1 Observations on the curve for the function $y = \ln x$

If someone starts moving from the origin towards +1 and looks downwards to observe the points on the curve $\ln x$ (refer Figure 3.10) he or she will initially get lost in a gorge like figure with unfathomable depth. As the observer comes closer and closer to $x = 1$ the gorge becomes more and more visible. When the observer reaches $x = 1$ a new terrain starts in the form of an inclined plane; as one keeps on moving to the right, the slope of the plane becomes less and less noticeable. But it never ceases to rise.

Suppose we walk k steps from the origin towards right and each step is equal to one unit. So at the end of our journey we will reach a height of $\ln k$. If we take one more step we will reach a height $\ln(k+1)$. So the increase in height is $\ln(k+1) - \ln k = \ln\frac{k+1}{k} = \ln(1 + \frac{1}{k})$. As k increases, $\frac{1}{k}$ approaches 0 and the increase in height approaches $\ln 1$

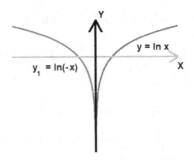

Figure 3.10: Graph of $y = \ln|x|$

i.e., zero. So, as we move more and more towards right, the rise of the slope becomes less and less visible.

3.8.2   Curve for the function $y = \ln(-x)$ and $y = \ln|x|$

One may hastily conclude that the function does not exist since negative numbers do not have logarithms. This conclusion is incorrect because of two reasons. Firstly, when we extend the number system by introducing complex numbers we find that logarithm of negative numbers also do exist. We will discuss this issue in greater details later in this chapter. Secondly, $-x$ does not always represent a negative number. When $x < 0, -x$ represents positive numbers. So it is obvious that the graph of the function $y = \ln(-x)$ for the value of $x < 0$ will be the mirror image of the graph for the function $y = \ln x$ and the former can be obtained easily by reflecting the later about the Y- axis.

We can combine the function $y = \ln x$ for $x \geq 0$ and the function $y = \ln -x$ for $x < 0$ to obtain a single function $y = \ln|x|$.

3.9   Leibnitz-Gregory Series for π

We know $e^{i\theta} = \cos\theta + i\sin\theta$

Let $\theta = \frac{\pi}{2}$; $\therefore$ $e^{i\frac{\pi}{2}} = \cos\frac{\pi}{2} + i.\sin\frac{\pi}{2} = i$

or, $\ln i = i\frac{\pi}{2}$

or, $i\frac{\pi}{2} = \ln\frac{i.(1-i)}{(1-i)} = \ln\frac{1-i^2)}{(1-i)} = \ln\frac{(1+i)}{(1-i)} = \ln(1+i) - \ln(1-i)$

We know, $\ln(1+x) = x - \frac{x^2}{2} + \frac{x^3}{3} - \frac{x^4}{4} + \cdots$

$$\therefore \ln(1+i) = i + \tfrac{1}{2} - \tfrac{1}{3}i - \tfrac{1}{4} + \tfrac{1}{5}i + \tfrac{1}{6} - \cdots$$

$$\therefore \ln(1-i) = -i + \tfrac{1}{2} + \tfrac{1}{3}i - \tfrac{1}{4} + \tfrac{1}{5}i + \tfrac{1}{6} - \cdots$$

$$\therefore i\tfrac{\pi}{2} = 2i - \tfrac{2}{3}i + \tfrac{2}{5}i - \tfrac{2}{7}i + \cdots$$

$$\therefore \tfrac{\pi}{4} = 1 - \tfrac{1}{3} + \tfrac{1}{5} - \tfrac{1}{7} + \cdots$$

It is reported in [15] that Nilakantha, an Indian Mathematician of fifteenth century also derived this equation.

While the above series looks beautiful it converges at a slow rate. By carrying out some modification in the above method Schellbach significantly increased the convergence rate.

3.10 Schellbach's modified series for π

$$
\begin{aligned}
i\frac{\pi}{2} &= \ln i = \ln\frac{(1+i)}{(1-i)} = \ln\frac{(5+5i)}{(5-5i)} = \ln\frac{(6+5i-1)}{(6-5i-1)} \\
&= \ln\frac{(2+i)(3+i)}{(2-i)(3-i)} = \ln\frac{(1+\tfrac{1}{2}i)(1+\tfrac{1}{3}i)}{(1-\tfrac{1}{2}i)(1-\tfrac{1}{3}i)}
\end{aligned}
$$

$$\text{or, } i\frac{\pi}{2} = \ln(1+\tfrac{1}{2}i) - \ln(1+\tfrac{1}{2}i) + \ln(1+\tfrac{1}{3}i) - \ln(1+\tfrac{1}{3}i)$$

Expanding the right hand side of the above equation we get,

$$\frac{\pi}{4} = 1.\left(\frac{1}{2}+\frac{1}{3}\right) - \frac{1}{3}.\left(\frac{1}{2^3}+\frac{1}{3^3}\right) + \frac{1}{5}.\left(\frac{1}{2^5}+\frac{1}{3^5}\right) - \frac{1}{7}.\left(\frac{1}{2^7}+\frac{1}{3^7}\right) + \cdots$$

This series converges at a much faster rate than Leibnitz-Gregory

Series. If we carefully observe the terms we notice that this series can be obtained from Leibnitz-Gregory Series by multiplying each term by a factor which is put inside a bracket and is less than one. These factors rapidly approach zero. This is the reason why it converges at a faster rate.

3.11 Logarithm as a limit of a product

We know, $e^h = 1 + h + \frac{h^2}{2!} + \frac{h^3}{3!} + \frac{h^4}{4!} + \cdots$

$$\therefore e^h - 1 = h + \frac{h^2}{2!} + \frac{h^3}{3!} + \frac{h^4}{4!} + \cdots$$

Figure 3.11: Dividing Z into n number of parts

$$\text{or,} \quad e^h - e^0 = h + \frac{h^2}{2!} + \frac{h^3}{3!} + \frac{h^4}{4!} + \cdots$$

$$\text{or,} \quad \frac{e^h - e^0}{h} = 1 + \frac{h}{2!} + \frac{h^2}{3!} + \frac{h^3}{4!} + \cdots$$

$$\text{or,} \quad \lim_{h \to 0} \frac{e^h - e^0}{h} = 1$$

Let $h = \frac{z}{n}$ (refer Figure 3.11)

$$\therefore \quad \lim_{n \to \infty} \frac{e^{\frac{z}{n}} - 1}{\frac{z}{n}} = 1$$

$$\text{or,} \quad \lim_{n \to \infty} \frac{n(\sqrt[n]{e^z} - 1)}{z} = 1$$

$$\text{or,} \quad \lim_{n \to \infty} n(\sqrt[n]{e^z} - 1) = z$$

Let $z = \ln x$

$$\text{or,} \quad \ln x = \lim_{n \to \infty} n(\sqrt[n]{x} - 1)$$

As $n \to \infty$, $\sqrt[n]{x} \to x^0 (= 1)$ $\quad \therefore$ when $n \to \infty$, $(\sqrt[n]{x} - 1) \to 0$

So, $\ln x$ is the limit of a product of two factors one of which n tends to infinity and other one $(\sqrt[n]{x} - 1)$ tends to zero.

Let us take $x = 2$. We know that $\ln 2 \approx 0.69314718$

Table 3.6: $\ln 2$ as the limit of a product

n	$(2^{\frac{1}{n}} - 1)$	$n.(2^{\frac{1}{n}} - 1)$
1	1	1
10	0.07177346	0.71773462
100	0.00695555	0.69555500
1000	0.00069339	0.69338746
10000	0.00006932	0.69317120
100000	0.00000693	0.69314958

In Table 3.6 we observe that as n increases, $(2^{\frac{1}{n}} - 1)$ decreases; however, the product of n and $(2^{\frac{1}{n}} - 1)$ comes closer and closer to $\ln 2$.

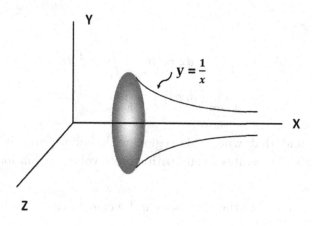

Figure 3.12: Torricelli's Trumpet

3.12 Torricelli's Trumpet

Earlier in this chapter we observed that the curve $y = \frac{1}{x}$ is associated with logarithmic and exponential functions. In the next chapter we will see that it actually represents a hyperbola. However, it has some other interesting features which will be demonstrated below.

If the region R (from 1 to ∞) under the curve $y = \frac{1}{x}$ is rotated about the x-axis a hollow solid resembling a trumpet or horn is formed. This is popularly known as **Torricelli's trumpet or Gabriel's horn** (refer Figure 3.12).

The volume is given by the improper integral

$$V = \int_1^\infty \pi . (\frac{1}{x})^2 . dx = \pi . \int_1^\infty x^{-2} . dx$$

This improper integral can now be determined as follows.

$$V = \pi . \lim_{b \to \infty} . \int_1^b x^{-2} . dx = \pi . \lim_{b \to \infty} (-\frac{1}{x}\big|_1^b)$$

$$= \pi . \lim_{b \to \infty} (1 - \frac{1}{b}) = \pi$$

Let us now find out the area A under the curve $\frac{1}{x}$ when x varies from 1 to ∞.

$$A = \int_1^\infty \frac{1}{x}.dx = \lim_{b\to\infty} . \int_1^b x^{-2}.dx$$

$$= \lim_{b\to\infty} \left(\ln x \, \Big|_1^b \right) = \lim_{b\to\infty} (\ln b) = \infty$$

So we find that when this region with infinite area is rotated about X-axis it generates a solid with a finite volume π (*an unexpected result!!*).

Surface area S of the trumpet can be calculated in the following way.

$$S = \int_1^\infty 2\pi \frac{1}{x}.dx = 2\pi \int_1^\infty \frac{1}{x}.dx = 2\pi A = \infty$$

So this trumpet can hold finite amount of liquid but it is not possible to colour its surface with finite amount of paint.

3.13   Logarithm of a complex number

In Chapter 2 we noticed that if the polar representation of a complex quantity is $1\angle\theta$ we can represent it as $e^{i\theta}$. However, if it has a magnitude r, we can express r in terms of the number e. Let us denote it as $r = e^\alpha$. In Figure 3.13 we have shown it as a straight line OP which has a magnitude of r (i.e., e^α) and which makes an angle θ with the real axis. So the complete representation of OP in terms of e will be

$$e^\alpha.e^{i\theta} = e^{\alpha+i\theta} = e^\alpha.(\cos\theta + i\sin\theta)$$
$$= e^\alpha \cos\theta + ie^\alpha \sin\theta$$
$$= x + iy = ON + iOM.$$

Since $e^{\alpha+i\theta} = x + iy$, we can write $\ln(x + iy) = \alpha + i\theta = \omega$ (a complex number). We may also call $x + iy$ as, another complex number z. $\therefore$ $\ln z = \omega$. The complex number ω has a real part $\alpha = \ln r = \ln|z|$ and an imaginary part $\theta = Arg(z)$.

However, if the line OP undergoes a rotation of 2π it will occupy the same position in the complex plane. As such, its rectangular representation will still be $x + iy$. Its representation in terms of number e will

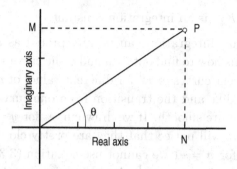

Figure 3.13: Point in a complex plane and corresponding number

be different and the exact representation will now be $e^{\alpha+i(\theta+2\pi)}$. If OP undergoes k number of 2π rotations the corresponding representation will be $e^{\alpha+i(\theta+k.2\pi)}$. Its rectangular representation in the complex plane will still be $x + iy$.

If we now treat the whole matter in a more general way we may write $\ln(x + iy) = \alpha + i(\theta + k \cdot 2\pi)$ where $k = 1, 2, 3, \cdots, \infty$ So, the logarithm of a complex number has infinite number of complex values where each of these resultant numbers will have same real part and their imaginary parts will differ by a factor 2π. The complex number $\alpha + i\theta$ is called the principal value of $\ln(x + iy)$.

Example:

$$\ln(4+i3) = \ln 5 + i.(\tan^{-1}\tfrac{3}{4} + 2k\pi) \qquad \text{where } k = 1, 2, 3, \cdots, \infty$$

3.14   Resolving an apparent contradiction

We are familiar with the following two equations.

$$\int x^n.dx = \frac{x^{n+1}}{n+1} + K_1 \quad \text{for} \quad n \neq -1 \tag{3.31}$$

K_1 is an integration constant.

$$\int \frac{1}{x}.dx = \int x^{-1}.dx = \ln x + K_2 \quad \text{for} \quad n = -1 \tag{3.32}$$

K_2 is an integration constant.

We know that integration can be interpreted as area. So equation (3.31) tells us how to find out area under the curve x^n when $n \neq -1$. If we draw different curves of x^n for different values of n we will notice that they look alike and the transition from one curve from another as n changes is quite smooth. If we draw curves for $y = x^{-0.9}$, $y = x^{-1}$ and $y = x^{-1.1}$ we will notice that they are pretty close to each other. In spite of this for $n = -1$ we cannot use equation (3.31) to calculate the area A under the curve x^{-1}. As such, we encounter an apparent contradiction which we will try to resolve here.

We will show that for values of n pretty close to -1 equation (3.31) becomes identical to equation (3.32).

$$A = \int_v^u \frac{1}{x}.dx = \ln v - \ln u = \ln \frac{v}{u} \quad \text{when } u, v > 0 \qquad (3.33)$$

We will now choose a value of $n = -1 + \epsilon$ where ϵ is a very small number and use equation (3.32).

$$A = \int_v^u x^{-1+\epsilon}.dx = \frac{v^\epsilon - u^\epsilon}{\epsilon} \quad \text{when} \quad \epsilon \neq 0 \qquad (3.34)$$

We know that $u = e^{\ln u}$ $\quad \therefore \quad u^\epsilon = e^{\epsilon \ln u}$

Similarly $v = e^{\ln v}$ $\quad \therefore \quad v^\epsilon = e^{\epsilon \ln v}$

$$\therefore \quad A = \frac{v^\epsilon - u^\epsilon}{\epsilon} = \frac{e^{\epsilon \ln v} - e^{\epsilon \ln u}}{\epsilon} \qquad (3.35)$$

However, we know that for small values of x, $e^x \approx 1 + x$

$$\therefore \quad A = \frac{e^{\epsilon \ln v} - e^{\epsilon \ln u}}{\epsilon} = \frac{(1 + \epsilon \ln v) - (1 + \epsilon \ln u)}{\epsilon}$$

$$= \frac{\epsilon(\ln v - \ln u)}{\epsilon} = \ln \frac{v}{u} \qquad (3.36)$$

From equation (3.35) and equation (3.36) it is clear that for small values of ϵ use of equation (3.31) and equation (3.32) yields same result.

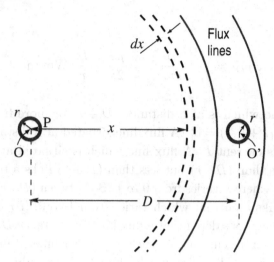

Figure 3.14: A two wire line and magnetic flux line

3.15 Applications

3.15.1 Inductance of a two wire line:

Let us consider a two wire line composed of hollow conductors of radius r with negligible thickness. At any instant of time current flows from one end to the receiving end through one line and returns through the other line. Now the centre of the left conductor (refer Figure 3.14) is considered as the origin O for our analysis. Let a current I ampere flows through the left conductor. Let the magnetic field intensity at a distance x meter from the origin be H_x.

$\therefore$ The mmf around the element is $2\pi x H_x = I$

or, $H_x = \frac{I}{2\pi x}$

$\therefore$ The flux density at a distance x, $B_x = \frac{\mu I}{2\pi x}$ Wb/m^2
where μ is the permeability of the medium.

Let us now consider the flux $d\phi$ in a thin tubular element of thickness dx at a distance x.

$$d\phi = \frac{\mu I}{2\pi x} dx \text{ Wb/m}$$

Now we will find out the total flux linkage between point P (on the outer surface of the left conductor) and O' (centre of right side conductor).

$$\psi_{PO'} = \int_r^D \frac{\mu I}{2\pi x}\, dx = \frac{\mu I}{2\pi} \ln \frac{D}{r} \text{ Wbt/m}$$

A flux line which is at a distance $(D + r)$ or more from O links zero current $[I + (-I) = 0]$. A flux line situated at a distance $(D - r)$ from O Links current I. A flux line which is situated at a distance which is more than $(D - r)$ but less than $(D + r)$ links a total current of $[I + k(-I)]$ where k varies from 0 to 1. So between $(D-r)$ to $(D+r)$ a flux line links a current which varies from 1 to 0. To simplify the problem we have considered that a flux line at a distance D links I and beyond D links zero current. As such, the upper limit of integration is set to D. Since any flux line inside the left side conductor does not link any current [since $I + (-I) = 0$] the lower limit of integration is set to r.

∴ The inductance of the left conductor $L_1 = \frac{\psi_{PO'}}{I_1} = \frac{\mu}{2\pi} \ln \frac{D}{r}$

Following the same method and noting that a current $(-I)$ as flows throgh the return conductor we can write down the expression for the inductance of the right side conductor as follows.

$$L_2 = \frac{\mu}{2\pi} \ln \frac{D}{r}$$

∴ The total inductance for the complete circuit $L = L_1 + L_2 = \frac{\mu}{\pi} \ln \frac{D}{r}$

3.15.2 Capacitance of a two wire line

To find out the capacitance between the two conductors of a two wire line we will first determine the Potential difference between two points a and b on the surface of the left and right conductors respectively (refer Figure 3.15) due to uniformly distributed charge of q coulombs/metre on the left conductor.

To start with let us consider that the left side conductor has a distributed charge of q per metre on it and assume the right side conductor is uncharged.

Using Gauss' theorem we can determine the magnitude of electric flux density D_x^f per metre at a point x metres as follows:

$$D_x^f = \frac{q}{2\pi x} \text{ coulomb/meter}^2$$

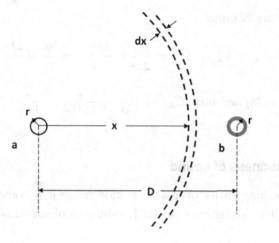

Figure 3.15: Capacitance calculation of a 2-wire line

The electric intensity at that point will be

$$E_x = \frac{q}{2\pi\epsilon x} \quad \text{where } \epsilon \text{ is the permittivity of the medium.}$$

The potential difference between points a and b due to the charged left side conductor will be

$$V_{ab}^L = \int_r^{D-r} E_x . dx = \int_r^{D-r} \frac{q}{2\pi\epsilon x} . dx = \frac{q}{2\pi\epsilon} \frac{D-r}{r} \quad \text{V}$$

Let us now determine the potential difference between two points due to uniformly distributed charge of $-q$ coulomb/metre on the right side conductor. Here we will assume that the left side conductor is uncharged. The potential difference between points b and a due to the charged right side conductor with a distributed charge of q coulomb per metre will be

$$V_{ba}^R = \int_r^{D-r} \frac{-q}{2\pi\epsilon x} . dx = \ln \frac{-q}{2\pi\epsilon} \frac{D-r}{r} \quad \text{V}$$

$$\therefore \quad V_{ab}^R = \frac{q}{2\pi\epsilon} \ln \frac{D-r}{r} \quad \text{V}$$

Potential difference V_{ab} between two points a and b when both the *conductors are charged:*

$$\therefore \; V_{ab} = V_{ab}^{L} + V_{ab}^{R} = \frac{2q}{2\pi\epsilon} \ln \frac{D-r}{r} = \frac{q}{\pi\epsilon} \ln \frac{D-r}{r} \; V$$

$$\therefore \; \text{The Capacitance } C_{ab} = \frac{q}{V_{ab}} = \frac{\pi\epsilon}{\ln \frac{D-r}{r}} \approx \frac{\pi\epsilon}{\ln \frac{D}{r}} \quad F/m \quad \text{since}$$
$$D \gg r$$

3.15.3 Loudness of sound

When something varies over a very wide range it becomes convenient to express it in logarithmic scale. Loudness L of sound is thus defined as follows.
$$L = \log_{10} \frac{I}{I_b} \text{ bell (after Alexander Graham Bell)}$$
Here I_b is the minimum intensity of sound that can be sensed by the human ear. If the intensity I of a sound is 10 times higher than I_b, then this sound will have a loudness L of 1 bell. If the intensity of a sound is 100 times higher than I_b, then its loudness will be 2 bell.

However, bell is a large unit; as such loudness L is usually measured in deci bell as follows.
$$L = 10 \log_{10} \frac{I}{I_b} \text{ db (deci bell)}$$

3.15.4 Magnitude of earthquake

Magnitude of earthquake is defined in Richter scale as follows.
$$R = \log_{10} \frac{A}{A_b}$$
Here A_b is the magnitude of the weakest earthquake that can be detected and recorded by the Seismograph and A is magnitude of the concerned earthquake.

3.15.5 Acidity of a substance

In Chemistry level of acidity of a substance is determined from the concentration of the hydrogen ion in moles per litre (x). However, this concentration varies over a very wide range for common substances. For Sodium Hydroxide it is 10^{-14}; for pure water it is 10^{-7}; for hydrochloric acid it is 1 or 10^{0}. If the concentration x is expressed as 10^{y} then $y \leq 0$. As such, to make the level of acidity (pH) always a positive number we define it as follows.
$$pH = -\log_{10} x$$

Pure water having a pH value of 7 is considered to be a neutral substance. Substances with pH values less than 7 are acidic in nature where as substances having pH values more than 7 are alkaline in nature.

3.16 Chronology of development of the concepts related to logarithm

- Works of the eminent Greek mathematician Archimedes (c. 287 – c. 212 B.C.) reveal that he had basic understanding of logarithms and their application. He, however, did not give any name to it.

- The mathematical works (the Jaina text named *Anuyoga Dwara Sutra*) by ancient Indian mathematicians around 200 AD reveal that they developed the laws of indices and they knew how to multiply numbers by adding exponents. Some historians are of the opinion that Jaina's developed logarithms to base 2.

- In eleventh century Ibon Jonuis, an Arab mathematician proposed a method of multiplication which can save computational labour significantly. The method is known as *Prosthaphaeresis*. The Greek word *Prosthesis* means addition and *aphaeresis* means subtraction.

- In sixteenth century two mathematicians, Paul Wittich and Christopher Clavius suggested similar methods using trigonometric tables to reduce computation during multiplication.

- In the late sixteenth century a lot of scientific work was going on in the field of long distance navigation, observational astronomy and science related to measurement and representation of earth. While working in these areas researchers had to frequently carry out multiplication and division of large numbers. As such, there was a pressing need to find a technique which could replace multiplication and division by easier processes like addition and subtraction.

- The Scottish mathematician John Napier made a pioneering contribution in this field. He introduced artificial numbers, named them as logarithms and used them to replace multiplication and division by addition and subtraction respectively. Napier's method is based on a kinematic framework. Historians of Mathematics are yet to understand fully his motivation behind this approach. He also invented an artefact known as Napier's bone to do the computational job in a mechanical way. He published two books in 1614 and 1617 (published after his death) on logarithm.

- Joost Burgi of Switzerland was contemporary of Napier. He was a clock maker by profession and a collaborator of Kepler. He also proposed methods which make correspondence between terms of an arithmetic series (red numbers) and those of geometric series (black numbers) and simplify computational work. Evidences reveal that he devised his method earlier than Napier's; however, he published his work later only in 1620.

- Henry Briggs, an eminent English mathematician of that age collaborated with Napier and made significant work to persuade the scientific community to accept logarithm. He dealt with powers of 10 and published a Table with eight decimals in 1617.

- In 1620, Edmund Grunter, an English mathematician, devised a stick which contained logarithm of numbers and used this device for calculation. William Oughtred first developed an useful slide rule which contained sliding parts with logarithmic scale on them. Many modifications came up later and slide rules were extensively used by the engineers and scientific workers more than three centuries till the advent of electronic calculators and computers in 1960s.

- Gregory of Saint Vincent highlighted in his published work in 1647 that for the curve $y = \frac{1}{x}$ if $x_0, x_1, x_2, x_3, \cdots$ are so chosen that

$$\int_{x_0}^{x_1} y.dx = \int_{x_1}^{x_2} y.dx = \int_{x_2}^{x_3} y.dx = \cdots$$

then $x_0, x_1, x_2, x_3, \cdots$ are in Geometric Progression.

Gregorie's student Alfons A. de Sarasa in 1649 finally interpreted the area under the curve $y = \frac{1}{x}$ as logarithm.

- Mercator also made use of the definition of the logarithm by means of the area under rectangular hyperbola. In his book and articles published in the late 1660s he argued that logarithm obtained in this manner differs from common logarithm with the base 10 only by a constant factor. He introduced the name *natural logarithm* or *hyperbolic logarithm* for the former logarithm. However, the greatest contribution of Mercator is the development of power series for the logarithm.

- Newton took up the ideas of Mercator and made two important contributions, namely, general binomial theorem and method of reversion of logarithmic series. By the latter he derived the series

$$1 + \frac{x}{1!} + \frac{x^2}{2!} + \frac{x^3}{3!} + \cdots \text{ (the series for } e^x)$$

- Euler introduced the idea of logarithmic function in his published works during the late 1760s and treat in a masterly way the integration of logarithmic functions although Leibnitz and Newton used the following relations earlier in their works.

$$\frac{d \ln x}{dx} = \frac{1}{x}; \qquad \int \frac{1}{x} . dx = \ln x$$

- At the beginning of seventeenth century Leibnitz and Johann Bernoulli vigorously debated on the existence and nature of logarithm of negative and imaginary quantities. Euler in 1749 concluded that logarithm of a number (positive, negative, imaginary and complex) has infinite number of values.

Bibliography

1. Courant, R. and Robbins, H., 1996 *What is Mathematics?* 2nd Edition (revised by Ian Stewart), Oxford University

2. Zeldovich, Ya. B. and Yaglom,I. M., 1987 *Higher Math for Beginners.* Mir Publishers, Moscow

3. Hall and Knight, 1891 *Higher Algebra* 4th Edition, MacMillan and Co.

4. Markushevich, A. I. 1981 *Areas and Logarithms* Mir Publishers, Moscow

5. Toeplitz, O., 2007 *The Calculus — A Genetic Approach* The University of Chicago Press, Chicago — London

6. Umbarger D. *Explaining Logarithms* (www.mathlogarithms.com)

7. Hogben L. *Mathematics for the Millions*

8. Clark, K. M. and Montelle, C, 2011 *Logarithms: the early history of a familiar function* Convergence (January, 2011), Mathematical Association of America

9. *Exponential and Logarithmic Functions — A Guide for Teachers* Australian Mathematical Sciences Institute

10. Stevenson, W.D., 1982 *Elements of Power System Analysis* 4th edition, MacGraw Hill Book Company

11. Roy, S., 2007 *Electrical Power Systems* Prentice Hall of India

12. Lefort, X. *History of Logarithms*

13. Klein, F. *Elementary Mathematics from an Advanced Stand Point*

14. Dorofeev, G., Potapov, M. and Rozov, N., 1973 *Elementary Mathematics — Selected Topics and Problem Solving* Mir Publishers, Moscow

15. Roy, R. *The Discovery of the Series Formula for π by Leibniz, Gregory and Nilakantha* Mathematics Mgazine, Vol. 63, No. 5(Dec.,1990), pp 291-306, Published by Mathematical Association of America

Chapter 4

Concept of Complex Angle and Hyperbolic Functions

"From the standpoint of electrical engineering arithmetic, the use of complex numbers and complex angles has vastly increased the power of computation. Without complex numbers and complex angles, the corresponding arithmetical solutions for the behaviour of electric lines and nets, would often be virtually unattainable. From the standpoint of applied mathematics, electrical engineering has entered two fields of basic, or so-called pure, mathematics; namely, complex numbers in higher algebra, and complex angles in higher trigonometry — branches which fifty years ago had little expectation of being generally used, and which were then known only to a few specialists — and has harnessed them into practical engineering service."

Dr. Arthur E. Kennelly in 1931

4.1 Introduction

Like numbers (refer Chapter-1) concept of angle has undergone many changes and extensions over a long historical period. The origin of the English word angle is a Latin word angulus which means corner. The intuitive idea about angle is associated with sharpness and bluntness of a corner. As such, the terms acute (Latin acutus means sharpened) angle and obtuse (Latin obtusus means blunted) angle have been coined. In ancient times Egyptians and Greeks defined angle as the opening between two straight lines at the point of their intersection on a plane. Usually this opening is made when one straight line rotates with respect to another over a circular path. People noticed that so long the rotation continues the length of the arc increases. If the straight line completes one full rotation the arc length will be equal to the circumference of the circle. The circumference of this circle is divided into 360 equal parts and each part is termed as one degree. One degree is further subdivided into sixty equal parts where each part is called a minute. Again one minute is subdivided into sixty equal parts where each part is called a second. Later on mathematicians defined angles in a more matured manner in the following ways.

1. Angle may be defined as the ratio of the length of the arc traversed during the rotation along a circular path to the length of the rotating straight line. (This ratio can be considered as the length of the arc in terms of the radius).

2. Angle may be defined as twice the area swept out by the straight line during rotation where the circle considered has unit radius.

3. Angle may be defined as the logarithm (base e) of the ratio of two co-planar vectors. Concept of imaginary angle and complex angle may then be introduced.

When we encounter complex angle its real part usually represents hyperbolic angle and imaginary part represents circular angle (In some problems, however, real part of a complex number represents circular angle and imaginary part represents hyperbolic angle). We know that a circular angle is produced when the tip of a straight line moves over a circle. Later in this chapter we will see that a hyperbolic angle is produced when the tip of a straight line moves over a hyperbola. Hyperbolic sine, cosine and other functions find wide applications in science and engineering. In the next section we will show that when a unit radius vector rotates in the counter clockwise direction the area traversed by it will be equal to half the circular angle.

4.2   Angle in terms of the area swept over during rotation

Let us take an infinitesimal arc with an arc length ds (refer Figure 4.1).

$$ds = r.d\theta$$

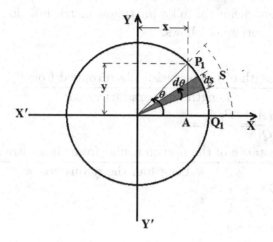

Figure 4.1: Circular rotation

$$\text{or,} \quad d\theta = \frac{ds}{r}$$

Integrating the above equation we get,

$$\text{or,} \quad \int_0^\theta d\theta = \theta = \frac{1}{r}\int_0^s ds = \frac{s}{r}$$

which expresses arc length interms of length of radius of the circle.
   If $r = 1$, we get,

$$\theta = s \tag{4.1}$$

The area of the sector is $dA = \frac{1}{2}r.ds = \frac{(r^2.d\theta)}{2}$

Integrating the above equation we get,

$$\int_0^A dA = \frac{r^2}{2}\int_0^\theta d\theta$$

$$A = \frac{r^2.\theta}{2} \tag{4.2}$$

When $r = 1$,

$$\theta = 2A \tag{4.3}$$

We may now define the following trigonometric functions associated with circular motion as follows.

$$\sin\theta \;=\; \frac{\text{Length of the perpendicular dropped from} P_1}{\text{Length of the radius vector}}$$

$$\;=\; \frac{P_1A}{OQ_1} = \frac{P_1A}{1} = y \tag{4.4}$$

$$\cos\theta \;=\; \frac{\text{Distance of the perpendicular from the centre of the circle}}{\text{Length of the radius vector}}$$

$$\;=\; \frac{OA}{OQ_1} = \frac{x}{1} = x \tag{4.5}$$

$$\tan\theta \;=\; \frac{\sin\theta}{\cos\theta} = \frac{y}{x} \tag{4.6}$$

Afterwards in this chapter we will start thinking in a more general way by considering the tip of a straight line moving along curves other than a circle. In [11] hyperbola and Bernoulli's lemniscate (a curve having the form of figure eight) have been considered. A hyperbolic angle is produced when the tip of a straight line moves over a hyperbola and a lemniscate angle is produced when the tip of a straight line moves over a Bernoulli's lemniscate. At present, however, we will consider only circular angle from a different view point.

4.3   Angle due to rotation from a vector view point and the concept of imaginary angle

Let us consider a vector OX_1 lying along the real axis (refer Figure 4.2). If it now rotates a little bit in the counter clock wise direction with O as centre it occupies a new position OX_2. The arc X_1X_2 is perpendicular to the radius vector OX_1. If the radius vector now again rotates a little bit in the counter clock wise direction occupying a new position OX_3 such that the arc length X_2X_3 is equal to the arc length X_1X_2 and arc X_2X_3 is perpendicular to the radius vector OX_2 we may write down the following equation for the incremental angle of rotation.

$$\text{Incremental angle} = \frac{X_1X_2}{OX_1} = \frac{\Delta x\angle 90^\circ}{x} = \Delta\theta\angle 90^\circ = i\Delta\theta$$

$$\text{Incremental angle} = \frac{X_2X_3}{OX_2} = \frac{\Delta x\angle(\Delta\theta+90^\circ)}{x\angle\Delta\theta} = \Delta\theta\angle 90^\circ = i\Delta\theta$$

where, arc length $X_1X_2 = $ arc length $X_2X_3 = \Delta x; |OX_1| = |OX_2| = x;$

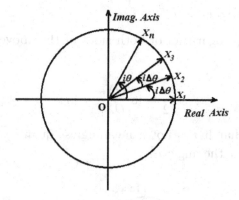

Figure 4.2: Stepped movement over a circle

$$|\Delta\theta| = \frac{|\Delta x|}{x}$$

We may now keep on rotating the radius vector in this stepped manner till the tip of the radius vector reaches the point X_n. It is important to note here that angle is now defined as the ratio of arc length to the radius vector where the radius vector keeps on changing at every step; first time it is OX_1, then OX_2, then OX_3 and so on. Finally the angle $\angle X_1 OX_n$ is termed as $i\theta$ since we can obtain it by adding up infinite numbers of $i\Delta\theta$. Using the notation of integral calculus we may now write

$$i\theta = \int_{OX_1}^{OX_n} \frac{dx}{x} = \ln OX_n - \ln OX_1 = \ln \frac{OX_n}{OX_1} \qquad (4.7)$$

Now let us consider three particular cases showing that logarithms of imaginary numbers, negative numbers and complex numbers are not absurd quantities but they represent circular angles or hyperbolic angle or both (refer Figure 4.3). We have already demonstrated these in Chapter 3; we will now show it again from a different view point. The discussion made below is based on the ideas expressed by Boyajian in [14].

CASE 1: When the arc $X_1 X_n$ is the quadrant of a circle X_n [refer Figure 4.3(a)] lies on the imaginary axis and *the circular angle is $\frac{\pi}{2}$* or the angle is $i\frac{\pi}{2}$. In the first narrative we use the adjective *circular* and in the second one we replace it by the symbol i. Here,

$$\frac{OX_n}{OX_1} = \frac{x\angle 90°)}{x\angle 0°} = 1\angle 90° = e^{i\frac{\pi}{2}} = i$$

Taking logarithm of both sides of the above equation we get,

$$\therefore \ \ i\frac{\pi}{2} = \ln\frac{OX_n}{OX_1} = \ln i \qquad\qquad (4.8)$$

So we find that $\ln i$ is not a meaningless quantity but it is the *circular angle* $\frac{\pi}{2}$ or the angle $i\frac{\pi}{2}$.

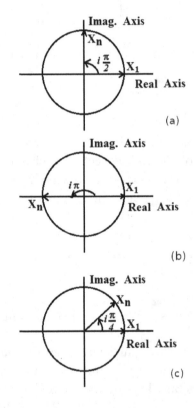

Figure 4.3: Rotational Angles expressed as Imaginary Angles

CASE 2: When X_n lies at the negative side of the real axis [refer Figure 4.3(b)] the angle is $i\pi$. At this point $\frac{OX_n}{OX_1} = \frac{x\angle 180°}{x\angle 0°} = 1\angle 180° = e^{\pi} = -1$

Taking logarithm of both sides of the above equation we get,

$$\therefore \quad i\pi = \ln \frac{OX_n}{OX_1} = \ln(-1) \tag{4.9}$$

So we find that $\ln(-1)$ is not an absurd quantity but its value is $i\pi$.

CASE 3: When the angle is $45°$ or $i\frac{\pi}{4}$ [refer Figure 4.3(c)]

$$\frac{OX_n}{OX_1} = \frac{x\angle 45°}{x\angle 0°} = 1\angle 45° = e^{i\frac{\pi}{4}} = \frac{1}{\sqrt{2}} + i\frac{1}{\sqrt{2}}$$

Taking logarithm of both sides of the above equation we get,

$$\therefore \quad i\frac{\pi}{4} = \ln \frac{OX_n}{OX_1} = \ln\left(\frac{1}{\sqrt{2}} + i\frac{1}{\sqrt{2}}\right) \tag{4.10}$$

So we note that logarithm of a complex number $(\frac{1}{\sqrt{2}} + i\frac{1}{\sqrt{2}})$ is $\frac{i\pi}{4}$.

4.4 Angle due to stretching or shrinking and the concept of real angle

As per the earlier definition of an angle it is the logarithm of a ratio of two lengths where numerator is in quadrature with the denominator. So we may now think about another possibility where the numerator is in the same direction as that of the denominator. Of course it would no longer define a circular angle but we may accept it as a new type of angle which is in quadrature with the circular angle thereby extending our notion of an angle. Traditionally it is called a hyperbolic angle. Later we will provide a justification of this name. Physically we may encounter it when a radius vector does not undergo any rotation but simply gets stretched (or shrunk) with the elapse of time (refer Figure 4.4).

Let us consider a vector OX_1 lying along the real axis. If it now stretches a little bit it becomes OX_2. The increment X_1X_2 is along the radius vector OX_1. If the radius vector now again stretches a little bit it becomes OX_3 such that the length X_2X_3 is equal to the arc length X_1X_2 and the increment X_2X_3 is along the radius vector OX_2.

Let us now consider the curve $y = \frac{1}{x}$ (refer Figure 4.5). We will consider points $X_1', X_2', \cdots, X_n'$ on this curve corresponding to points $X_1, X_2, \cdots, X_n$ on the X-axis.

So the abscissa of the point X_1' is OX_1 and the ordinate is $X_1'X_1 = \frac{1}{OX_1}$.

Similarly the abscissa of the point X_2' is OX_2 and the ordinate is $X_2'X_2 = \frac{1}{OX_2}$ and so on.

we may write down the following equation for the incremental angles due to stretching.

Incremental angle (hyperbolic)

$$
\begin{aligned}
\Delta\alpha_1 &= \frac{X_1 X_2}{OX_1} = X_1 X_2 . X_1 X_1' = \text{Area of } X_1 X_2 X_2'' X_1' \\
&= \text{Area of } X_1 X_2 X_2' X_1' + \epsilon_1 \\
&= \ln\frac{OX_2}{OX_1} + \epsilon_1
\end{aligned}
$$

Here ϵ_1 is the area of the triangle like figure $X_1' X_2' X_2''$ which introduces error in the calculation of incremental angle $\Delta\alpha_1$.

If the step size is very small X_2 will be very near to X_1 and error ϵ_1 will be negligibly small.

$$
\therefore \ \Delta\alpha_1 = \ln\frac{OX_2}{OX_1}
$$

$$
\text{Similarly,} \ \ \Delta\alpha_2 = \ln\frac{OX_3}{OX_2} \ \ \text{and so on.}
$$

For a very large value of n,

$$
\alpha = \Sigma_{j=1}^{j=n}\alpha_j = \ln\frac{OX_n}{OX_1} \tag{4.11}
$$

Since $y = \frac{1}{x}$ represents an hyperbolic curve and α is the area under this curve it is called an hyperbolic angle. In section 4.10 it will be shown that this area will be same as the area swept over $(OX_1' X_n')$ when a point moves over the curve $y = \frac{1}{x}$ from X_1' to X_n'. In section 4.8 it will be shown that this form of hyperbola is a transformed version of the standard form of hyperbola $x^2 - y^2 = 1$.

It is important to point out here that $\angle\frac{OX_n'}{OX_1'}$ is not the hyperbolic angle α.

4.5 Complex Angle

When a radius vector simultaneously undergoes both rotation and stretching (or shrinking) with the elapse of time we come across

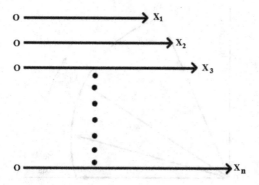

Figure 4.4: Stretching of a vector in stepped manner

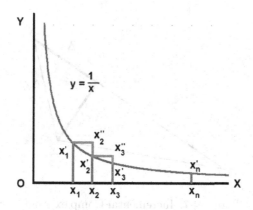

Figure 4.5: Stretching and the area under the curve $y = \frac{1}{x}$

an angle which has both a hyperbolic (real) part α and a circular (imaginary) part $i\theta$; as such, we get a complex angle $\theta' = \alpha + i\theta$.

Let us assume that the tip of a vector is initially at position OX_1 (refer Figure 4.6); then it moves to OX_2, $OX_3 \cdots$ and finally it reaches at OX_n. It is important to note that while the position vector takes different positions OX_1, OX_2, OX_3 it undergoes rotation as well as elongation since $|OX_1| < |OX_2| < |OX_3|$.

Referring Figure 4.7 we now write $\Delta x = \Delta x' + i.\Delta x''$

$$\therefore \quad \theta' = \alpha + i\theta = \int_{OX_1}^{OX_n} \frac{dx}{x} = \ln \frac{OX_n}{OX_1} \qquad (4.12)$$

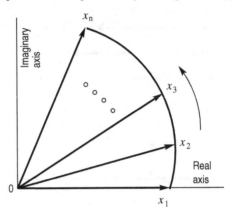

Figure 4.6: A planar vector undergoing both rotation and stretching

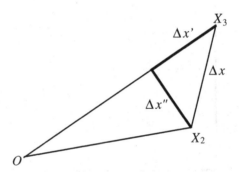

Figure 4.7: Incremental Complex Angle

It is important to note that we now come across a new definition of angle. *According to this new definition both the hyperbolic angle α and the circular angle $i\theta$ are sums of infinite number of per unit changes and they have directions as well.* While calculating α the change in length of the position vector in the radial direction is considered; while calculating $i\theta$ the change of the position vector in the tangential direction is considered. A rotating axis which coincides with the radial direction at each instant is considered as the real axis and the corresponding rotating imaginary axis coincides with its tangent at each instant. *As such the circular angle $i\theta$ is 90° ahead of the hyperbolic angle α. The complex angle $\theta' = \alpha + i\theta$ combines per unit changes in both the directions.*

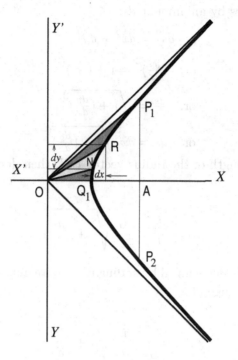

Figure 4.8: Hyperbola $x^2 - y^2 = 1$

From equation (4.12) we get

$$\frac{OX_n}{OX_1} = e^{\theta'} = e^{\alpha+i\theta} = e^{\alpha}.e^{i\theta}$$

$$\therefore \ OX_n = OX_1.e^{\alpha}.e^{i\theta} \qquad (4.13)$$

It is important to note that the exponent of exponential (e) repre-sents an angle; if the exponent is imaginary the angle is a circular one and if it is a real number the angle is hyperbolic.

4.6 Hyperbolic angle and Hyperbolic Functions for a hyperbola $x^2 - y^2 = 1$

Let us consider the unit hyperbola $x^2 - y^2 = a^2 = 1$ (4.14)

If ds represents an infinitesimal length of arc (refer Figure 4.8)such that when x changes by dx, y changes by an amount dy and the arc length s changes by an amount ds.

$$\therefore \quad ds^2 = dx^2 + dy^2$$

$$\text{or,} \quad \left(\frac{ds}{dx}\right)^2 = 1 + \left(\frac{dy}{dx}\right)^2$$

$$\text{or,} \quad \frac{ds}{dx} = \sqrt{1 + \left(\frac{dy}{dx}\right)^2}$$

$$\text{or,} \quad ds = dx.\sqrt{1 + \left(\frac{dy}{dx}\right)^2}$$

If l is the length of the radius vector ON, then denoting $\frac{ds}{l} = d\alpha$ we can write

$$d\alpha = \frac{ds}{l} = \frac{dx}{l}.\sqrt{1 + \left(\frac{dy}{dx}\right)^2} \tag{4.15}$$

α can now be determined by adding up infinite number of $\Delta\alpha = \frac{\Delta s}{l}$ present in the range $(1, x)$.

$$\therefore \quad \alpha = \int_1^x \frac{dx}{l}.\sqrt{1 + \left(\frac{dy}{dx}\right)^2} \tag{4.16}$$

The function α is the integrated ratio of the length of an infinitesimal arc to the length of the radius vector l at any point R.

$$\therefore \quad \alpha = \int \frac{\text{Length of arc}}{\text{Length of radius}}. \tag{4.17}$$

by analogy to the circular angle, the function α is called an hyperbolic angle. It is important to note here that in case of a purely circular motion the length of radius r remains constant(its angle, however, keeps on changing) while the movement over a hyperbola causes length of radius vector l to vary. Like the angle traversed in case of a circular motion it can be shown that the function α is equal to the twice the area of the hyperbolic sector (the proof will be given later in this chapter).

Now, $l = \sqrt{x^2 + y^2}$. From equation (4.14), we know

$$y^2 = x^2 - 1 \tag{4.18}$$

Differentiating the above equation we get,

$$2x.dx = 2y.dy$$

$$\text{or,} \quad \frac{dy}{dx} = \frac{x}{y} \tag{4.19}$$

Substituting this in equation (4.16) we get,

$$\alpha = \int_1^x \frac{dx}{\sqrt{x^2 + y^2}} \cdot \frac{\sqrt{x^2 + y^2}}{y}$$

$$\therefore \quad \alpha = \int_1^x \frac{dx}{\sqrt{x^2 - 1}} = [\ln(x + \sqrt{x^2 - 1})]_1^x = \ln(x + \sqrt{x^2 - 1})$$

$$\therefore \quad e^\alpha = x + \sqrt{x^2 - 1} \tag{4.20}$$

Squaring both sides we get,

$$e^{2\alpha} = x^2 + x^2 - 1 + 2.x.\sqrt{x^2 - 1}$$

$$\text{or,} \quad e^{2\alpha} + 1 = 2x(x + \sqrt{x^2 - 1}) = 2x.e^\alpha$$

$$\text{or,} \quad x = \frac{e^{2\alpha} + 1}{2.e^\alpha} = \frac{e^\alpha + e^{-\alpha}}{2} \tag{4.21}$$

$\because y^2 = x^2 - 1$ substituting the value of x we get

$$y^2 = \frac{(e^\alpha + e^{-\alpha})^2}{4} - 1$$

$$= \frac{e^{2\alpha} + e^{-2\alpha} + 2 - 4}{4}$$

$$= \frac{e^{2\alpha} + e^{-2\alpha} - 2.e^\alpha.e^{-\alpha}}{2^2}$$

$$\therefore \quad y = \frac{e^\alpha - e^{-\alpha}}{2} \tag{4.22}$$

$$\therefore \quad \frac{y}{x} = \frac{e^\alpha - e^{-\alpha}}{e^\alpha + e^{-\alpha}} \tag{4.23}$$

As we earlier defined trigonometric functions in case of circular motion we will define hyperbolic functions in a similar way.

$$\sinh \alpha = \frac{\text{Length of the perpendicular dropped from } P_1}{\text{Initial length of the Radius vector}} = \frac{P_1 A}{OQ_1} = \frac{y}{1}$$

$$\therefore \quad \sinh \alpha = \frac{e^\alpha - e^{-\alpha}}{2} \tag{4.24}$$

$$\cosh \alpha = \frac{\text{Perpendicular distance of } P_1 A \text{ from the origin}}{\text{Initial length of the Radius vector}} = \frac{OA}{OQ_1} = \frac{x}{1}$$

$$\therefore \quad \cosh \alpha = \frac{e^\alpha + e^{-\alpha}}{2} \qquad (4.25)$$

Dividing equation (4.24) by equation (4.25) we get

$$\tanh \alpha = \frac{y}{x} = \frac{e^\alpha - e^{-\alpha}}{e^\alpha + e^{-\alpha}} \qquad (4.26)$$

In case of a unit circle any point on its periphery has an abscissa of $\cos \theta$ and ordinate of $\sin \theta$, where θ is the circular angle. In case of a unit equilateral hyperbola any point on its periphery has an abscissa of $\cosh \alpha$ and ordinate of $\sinh \alpha$, where α is the hyperbolic angle.

It is important to note that

$$x^2 - y^2 = \cosh^2 \alpha - \sinh^2 \alpha = \frac{(e^\alpha + e^{-\alpha})^2}{4} - \frac{(e^\alpha - e^{-\alpha})^2}{4} = \frac{2+2}{4} = 1$$

4.7   Area swept by a straight line joining the origin and a point moving over a hyperbola $x^2 - y^2 = 1$

We will consider a small horizontal rectangle of height dy (refer Figure 4.9) and its base connects two points with abscissa x' and x where the first point lies on the straight line $y = (\tanh \alpha).x$ and second point lies on the hyperbola; both have the same ordinate y. We will now find the following area by integration.

$$A = \int_0^{\sinh \alpha} (x' - x).dy$$

Since x' is the abscissa of a point lying on hyperbola $x' = \sqrt{y^2 + 1}$; since x is the abscissa of a point lying on the straight line $y = \tanh \alpha$, $xx = \frac{\cosh \alpha}{\sinh \alpha}.y$

$$\therefore \quad A = \int_0^{\sinh \alpha} [\sqrt{y^2 + 1} - \frac{\cosh \alpha}{\sinh \alpha}.y].dy$$

$$= \int_0^{\sinh \alpha} \sqrt{y^2 + 1}.dy - \int_0^{\sinh \alpha} \frac{\cosh \alpha}{\sinh \alpha}.y.dy$$

$$\text{or, } A = A_1 - A_2 \qquad (4.27)$$

where

$$A_1 = \int_0^{\sinh \alpha} \sqrt{y^2 + 1}.dy \quad \text{and} \quad A_2 = \int_0^{\sinh \alpha} \frac{\cosh \alpha}{\sinh \alpha}.y.dy$$

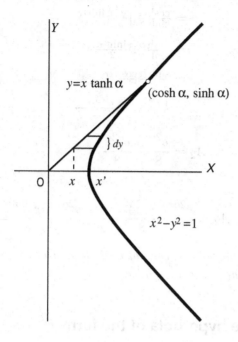

Figure 4.9: Area swept by a line whose end point moves over a hyperbola

Let $y = \sinh u$ $\therefore$ $dy = \cosh u . du$ and $\sqrt{1 + y^2} = \cosh u$

When $y = 0$, $\sinh u = 0$, $\therefore$ $u = 0$

When $y = \sinh \alpha$, $u = \alpha$

$$\therefore A_1 = \int_0^\alpha \cosh u . \cosh u . du$$

$$= \int_0^\alpha \cosh^2 u . du$$

$$= \int_0^\alpha \frac{(e^u + e^{-u})^2}{4} . du$$

$$= \tfrac{1}{4} \int_0^\alpha (e^{2u} + e^{-2u} + 2) . du$$

$$= \tfrac{1}{4} . \left(\frac{e^{2\alpha}}{2} - \frac{e^{-2\alpha}}{2} + 2\alpha \right)$$

$$\therefore A_1 = \frac{e^{2\alpha}}{8} - \frac{e^{-2\alpha}}{8} + \frac{\alpha}{2} \qquad (4.28)$$

$$A_2 = \frac{\cosh \alpha}{\sinh \alpha} \int_0^{\sinh \alpha} y . dy$$

$$= \frac{\cosh \alpha}{\sinh \alpha} \cdot \left| \frac{y^2}{2} \right|_0^{\sinh \alpha}$$

$$= \frac{\cosh \alpha}{\sinh \alpha} \cdot \frac{\sinh^2 \alpha}{2}$$

$$= \frac{\sinh \alpha \cosh \alpha}{2}$$

$$= \frac{1}{2} \frac{e^\alpha - e^{-\alpha}}{2} \cdot \frac{e^\alpha + e^{-\alpha}}{2}$$

$$\therefore \quad A_2 = \frac{e^{2\alpha}}{8} - \frac{e^{-2\alpha}}{8} \tag{4.29}$$

$$\therefore \quad A = A_1 - A_2 = \frac{e^{2\alpha}}{8} - \frac{e^{-2\alpha}}{8} + \frac{\alpha}{2} - \frac{e^{2\alpha}}{8} + \frac{e^{-2\alpha}}{8} = \frac{\alpha}{2} \tag{4.30}$$

So we note that both in case of a circle and in case of a hyperbola the area is half the angle subtended.

4.8   From the hyperbola of the form $x^2 - y^2 = 1$ to the hyperbola of the form $u.v = 1$

The equation of an equilateral hyperbola is $x^2 - y^2 = 1$.

$\therefore \quad (x + y).(x - y) = 1$

Now let $u = x + y$ and $v = x - y$; (refer Figure 4.10) $\therefore \quad u.v = 1$

$$\therefore \quad u = \frac{e^\alpha + e^{-\alpha}}{2} + \frac{e^\alpha - e^{-\alpha}}{2} = e^\alpha \tag{4.31}$$

$$\text{and} \quad v = \frac{e^\alpha + e^{-\alpha}}{2} - \frac{e^\alpha - e^{-\alpha}}{2} = e^{-\alpha} \tag{4.32}$$

From the above we also obtain

$$x = \frac{(u + v)}{2} \tag{4.33}$$

$$\text{and} \quad y = \frac{u - v}{2} \tag{4.34}$$

In Figure 4.10 points A and B lie on the curve $x^2 - y^2 = 1$ and A' and B' are the corresponding points lying on the curve $u.v = 1$.

We will now show that a magnification of $\sqrt{2}$ occurs in the length of a line when we move from $X - Y$ plane to $U - V$ plane.

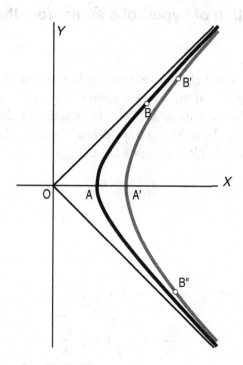

Figure 4.10: From $x^2 - y^2 = 1$ to $u.v = 1$

Let $P_1(x_1, y_1)$ and $Q_1(x_2, y_2)$ are two points in $X - Y$ plane and the corresponding points in the $U - V$ plane are $P(u_1, v_1)$ and $Q(u_2, v_2)$ respectively.

Length of the straight line $P_1 Q_1$ is $\sqrt{(x_2 - x_1)^2 + (y_2 - y_1)^2)}$

Length of $PQ = \sqrt{(u_2 - u_1)^2 + (v_2 - v_1)^2)}$

$$= \sqrt{(x_2 + y_2 - x_1 - y_1)^2 + (x_2 - y_2 - x_1 + y_1)^2}$$

$$= \sqrt{[(x_2 - x_1) + (y_2 - y_1)]^2 + [(x_2 - x_1) - (y_2 - y_1)]^2}$$

$$= \sqrt{2(x_2 - x_1)^2 + 2(y_2 - y_1)^2}$$

$$= \sqrt{2}\sqrt{(x_2 - x_1)^2 + (y_2 - y_1)^2} = \sqrt{2}P_1 Q_1$$

$\therefore$ $P_1 Q_1$ of $X - Y$ plane gets magnified by $\sqrt{2}$ and becomes straight line PQ in the $U - V$ plane.

$\therefore$ *Any area in X-Y co-ordinate system will be magnified by $\sqrt{2} \times \sqrt{2} = 2$ when it is transferred in U-V co-ordinate system.*

4.9 Calculation of hyperbolic angle from the curve $v = \frac{1}{u}$

When the tip of a straight line originating from point O starts moving over the curve $v = \frac{1}{u}$ it traverses a curved length ds in a very small time interval. During this movement the length l of the straight line keeps on changing. In section 4.6 we defined incremental angle $d\alpha$ in the following manner (Refer equation (4.15))

$$d\alpha = \frac{ds}{l}$$

We know,

$$ds^2 = du^2 + dv^2$$
$$\text{or, } \left(\frac{ds}{du}\right)^2 = 1 + \left(\frac{dv}{du}\right)^2$$
$$\text{or, } \frac{ds}{du} = \sqrt{1 + \left(\frac{dv}{du}\right)^2}$$
$$\text{or, } ds = du.\sqrt{1 + \left(\frac{dv}{du}\right)^2}$$

Again, $l = \sqrt{u^2 + v^2} = \sqrt{u^2 + \frac{1}{u^2}} = \frac{\sqrt{1+u^4}}{u}$

$$\frac{dv}{du} = -\frac{1}{u^2}$$
$$\therefore \quad d\alpha = \frac{ds}{l} = \frac{du}{l}\sqrt{1 + \left(\frac{dv}{du}\right)^2} = \frac{du}{u}$$

Let us now consider Figure 4.11 where $OQ' = 1$ and $OP' = u$. So the hyperbolic angle α traversed while the tip of the straight line moves from Q to P can be computed as follows.

$$\alpha = \int_1^u \frac{1}{u}.du = \ln u - \ln 1 = \ln u \qquad (4.35)$$

It is important to note that change in u i.e., du takes place in the direction of u. As such the incremental hyperbolic angle $d\alpha$ is a real

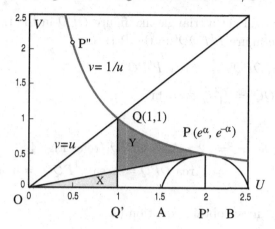

Figure 4.11: Tip of a line moving over the curve $v = \frac{1}{u}$

quantity. Since we get the hyperbolic angle α by integrating $d\alpha$ the former is also a real quantity.

We can also write $u = e^\alpha$. As such, we notice that real exponent of e represents an hyperbolic angle.

In Figure 4.11

$$\cosh \alpha = \frac{e^\alpha + e^{-\alpha}}{2} = \frac{OP' + P'B}{2} = \frac{OB}{2}$$

$$\sinh \alpha = \frac{e^\alpha - e^{-\alpha}}{2} = \frac{OP' - P'A}{2} = \frac{OA}{2}$$

$$\tanh \alpha = \frac{\sinh \alpha}{\cosh \alpha} = \frac{OA}{OB}$$

4.10 Calculation of traversed area while the tip of a straight line moves over the curve $v = \frac{1}{u}$

As we move from Q to P (refer Figure 4.11) a triangle like area OPQ is generated. Here $\angle OQ'Q = 90°$ and $OP'P = 90°$

Area $\triangle OQQ' = \frac{1}{2}.OQ'.Q'Q = \frac{1}{2}.1.1 = \frac{1}{2}$

Area $\triangle OPP' = \frac{1}{2}.OP'.P'P = \frac{1}{2}.u.u = \frac{1}{2}$

$\therefore$ Area $\triangle OQQ'$ = Area $\triangle OPP'$

Now from both sides first subtract area X and then add area Y.

As a result we will get a triangle like figure OPQ in the L.H.S. and a trapezium like figure $P'PQQ'$ in the R.H.S.

$$\therefore \quad \text{Area } OPQ \;=\; \text{Area } P'PQQ'$$

Area $P'PQQ' = \int_1^u \frac{1}{u}.du = \ln u$

$$\text{Angle } \alpha \;=\; 2 \times \text{Area } OP_1 Q_1 (\text{refer Fig. } 4.8)$$
$$=\; \text{Area } OPQ = \text{Area } P'PQQ' = \ln u$$

So once again we obtain equation (4.35).

4.11   Trigonometric functions of imaginary variable and Hyperbolic functions

We know that $\cos\theta = \frac{e^{i\theta}+e^{-i\theta}}{2}$ and $\sin\theta = \frac{e^{i\theta}-e^{-i\theta}}{2i}$

$$\text{Let } \theta \;=\; i\phi$$
$$\therefore \quad \cos(i\phi) \;=\; \frac{e^{-\phi} + e^{\phi}}{2}$$
$$\text{and } \sin(i\phi) \;=\; \frac{e^{-\phi} - e^{\phi}}{2i} = -\frac{e^{\phi} - e^{-\phi}}{2i} = i.\frac{e^{\phi} - e^{-\phi}}{2}$$

But we know that $\cosh\phi = \frac{e^{\phi}+e^{-\phi}}{2}$, $\sinh\phi = \frac{e^{\phi}-e^{-\phi}}{2}$

$$\therefore \quad \cos(i\phi) = \cosh\phi \qquad (4.36)$$
$$\text{and } \quad \sin(i\phi) = i.\sinh\phi \qquad (4.37)$$

4.12   Trigonometric and Hyperbolic functions of complex angle $\alpha + i\beta$

A point P in a complex plane may be represented by $e^{\alpha+i\beta} = e^{\alpha}.e^{i\beta}$. Here the length of the line OP connecting the point P to the origin $O = e^{\alpha}$ (Refer Fig. 3.14 of Chapter 3). OP makes a circular angle β with the real axis.

We may now rewrite $e^{\alpha+i\beta}$ as follows.

$$e^{\alpha+i\beta} = e^{i(\beta-i\alpha)}$$

$$\therefore \quad \cos(\beta-i\alpha) = \cos\beta\cos i\alpha + \sin\beta\sin i\alpha = \cos\beta\cosh\alpha + i\sin\beta\sinh\alpha \tag{4.38}$$

Similarly,

$$\therefore \quad \sin(\beta-i\alpha) = \sin\beta\cos i\alpha - \cos\beta\sin i\alpha = \sin\beta\cosh\alpha - i\cos\beta\sinh\alpha \tag{4.39}$$

equation (4.38) and equation (4.39) show that *trigonometric functions of a complex angle are complex quantities of the form* $a \pm ib$.

Now we will write similar expressions for hyperbolic sine and cosine functions.

$$\cosh(\alpha + i\beta) = \cosh\alpha\cosh i\beta + \sinh\alpha\sinh i\beta$$

Now,

$$\cosh i\beta = \frac{e^{i\beta} + e^{-i\beta}}{2} = \cos\beta \quad \text{and} \quad \sinh i\beta = \frac{e^{i\beta} - e^{-i\beta}}{2} = i\sin\beta$$

$$\therefore \quad \cosh(\alpha + i\beta) = \cosh\alpha\cos\beta + i\sinh\alpha\sin\beta \tag{4.40}$$

Again,

$$\begin{aligned} \sinh(\alpha + i\beta) &= \sinh\alpha\cosh i\beta + \cosh\alpha\sinh i\beta \\ &= \sinh\alpha\cos\beta + i\cosh\alpha\sin\beta \end{aligned} \tag{4.41}$$

Equation (4.40) and equation (4.41) show that *hyperbolic functions of a complex angle are also complex quantities of the form* $a \pm ib$.

4.13 Applications

Many natural and physical processes can be described by differential equations; very often we find that functions like $A_1 e^{mt} + A_2 e^{-mt}$ satisfy those equations where A_1 and A_2 are constants. Here m may be a real number or an imaginary number or a complex number. Two exponential functions with real exponents may produce a hyperbolic function indicating increase or decrease in the magnitude of a certain quantity. Two exponential functions with imaginary exponents may

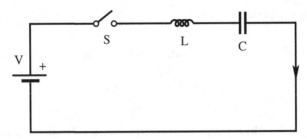

Figure 4.12: Switching on an LC circuit

produce an oscillation which can be mathematically described by cosine (or sine) type of circular functions. Two exponential functions with complex conjugate numbers as their exponents may produce a rising or decaying oscillation.

4.13.1   A DC Voltage Source (V) is switched on to an LC circuit

We now refer to Figure 4.12 and write down the following equation.

$$L.\frac{di_c}{dt} + \frac{\int i_c.dt}{C} = V \tag{4.42}$$

Differentiating equation (4.42) we get,

$$L.\frac{d^2 i_c}{dt^2} + \frac{i_c}{C} = 0$$

If we denote $s = \frac{d}{dt}$ we may write

$$(Ls^2 + 1/C)i_c = 0$$

$$\therefore \ s^2 + \frac{1}{LC} = 0$$

$$\text{or, } \ s_{1,2} = \pm\frac{1}{\sqrt{LC}} = \pm i\omega_n$$

$$\therefore \ i_c = A_1 e^{i\omega_n t} + A_2 e^{-i\omega_n t} \tag{4.43}$$

where A_1 and A_2 are constants.

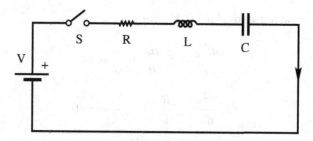

Figure 4.13: Switching on a RLC circuit

Since current can not change abruptly at $t = 0$, $i_c = 0$.

$$\therefore \ A_1 + A_2 = 0$$

$$\therefore \ i_c = A_1 e^{i\omega_n t} - A_1 e^{-i\omega_n t}$$

and $\dfrac{di_c}{dt} = A_1(i\omega_n e^{i\omega_n t} + i\omega_n e^{-i\omega_n t}) = 2A_1.i\omega_n.cos\omega_n t$

Since voltage across the capacitor can not change abruptly, at $t = 0$ voltage across the capacitor will remain zero and a voltage V will appear across the inductor L.

$$\therefore \ L.\dfrac{di_c}{dt} = V$$

$$\therefore \ L.2A_1.i\omega_n = V$$

$$\text{or,} \ A_1 = \dfrac{V}{2i\omega_n L}$$

$$\therefore \ i_c = \dfrac{V}{\omega_n L}.\dfrac{(e^{i\omega_n t} - e^{-i\omega_n t})}{2i} = \dfrac{V}{\omega_n L}\sin\omega_n t \qquad (4.44)$$

4.13.2   A DC Voltage Source is switched on to a RLC circuit

We now refer to Figure 4.13 and write down the following equation.

$$Ri_c + L.\dfrac{di_c}{dt} + \dfrac{\int i_c.dt}{C} = V \qquad (4.45)$$

Differentiating equation (4.45) we get

$$L.\frac{d^2 i_c}{dt^2} + R\frac{di_c}{dt} + \frac{i_c}{C} = 0$$

$$\text{or,} \quad \frac{d^2 i_c}{dt^2} + \frac{R}{L}\frac{di_c}{dt} + \frac{i_c}{LC} = 0$$

If we denote $s = d/dt$ we may write

$$s^2 + \frac{R}{L}.s + \frac{i_c}{LC} = 0$$

Solving the above quadratic equation we get following two solutions,

$$s_1 = \frac{\frac{-R}{L} + \sqrt{(\frac{R}{L})^2 - \frac{4}{LC}}}{2} = \frac{-R}{2L} + \sqrt{(\frac{R}{2L})^2 - \frac{1}{LC}} = -\alpha + \beta$$

$$s_2 = \frac{\frac{-R}{L} - \sqrt{(\frac{R}{L})^2 - \frac{4}{LC}}}{2} = \frac{-R}{2L} - \sqrt{(\frac{R}{2L})^2 - \frac{1}{LC}} = -\alpha - \beta$$

Here $\alpha = \frac{R}{2L}$; $\beta = \sqrt{(\frac{R}{2L})^2 - \frac{1}{LC}} = \sqrt{\alpha^2 - \omega_n^2}$ and $\omega_n = \frac{1}{\sqrt{LC}}$

The radicand of β may be positive, zero or negative. For positive radicand roots will be negative real and distinct; for zero radicand roots will be negative real and repeated; and for negative radicand roots will be complex conjugate. We will consider the two cases where roots are distinct (not repeated).

Case I: Roots are negative real and distinct

When $(\frac{R}{2L})^2 > \frac{1}{LC}$, β is a positive real number and $\beta < \alpha$.
Now,
$$i_c = C_1 e^{(-\alpha+\beta)t} + C_2 e^{(-\alpha-\beta)t}$$

where C_1, C_2 are constants.

Because of the presence of L, current can not change instantaneously.

∴ At $t = 0$, $i_c = 0$

As such, $C_1 + C_2 = 0$ ∴ $C_2 = -C_1$

$$\therefore \quad i_c = C_1(e^{(-\alpha+\beta)t} - e^{(-\alpha-\beta)t}) \tag{4.46}$$

Since the voltage across the capacitor can not change abruptly at

$t = 0$ it will remain zero.

More over, at $t = 0$, voltage across the resistance $i_c R$ will remain zero since at $t = 0$ current will remain zero.

So at $t = 0$, voltage across the inductor $L\frac{di_c}{dt} = V$

Now $\frac{di_c}{dt} = C_1[(-\alpha + \beta)e^{(-\alpha+\beta)t} - (-\alpha - \beta)e^{(-\alpha-\beta)t}]$

$\therefore$ At $t = 0, \frac{V}{L} = C_1(-\alpha + \beta + \alpha + \beta)$

$\therefore$ $C_1 = \frac{V}{2\beta L}$

$\therefore$ $i_c = \frac{V}{2\beta L}[e^{(-\alpha+\beta)t} - e^{(-\alpha-\beta)t}] = \frac{V}{2\beta L}[e^{-\alpha t}.e^{-\beta t} - e^{-\alpha t}.e^{-\beta t}]$

$\quad = \frac{V}{\beta L}e^{-\alpha t}\frac{e^{\beta t}-e^{-\beta t}}{2}$

$$\therefore \quad i_c = \frac{V}{\beta L}e^{-\alpha t}\sinh\beta t \tag{4.47}$$

Case II: Complex conjugate roots

When $(\frac{R}{2L})^2 < \frac{1}{LC}$, β is an imaginary number;

Let $\quad \beta = j\gamma$ where $\quad \gamma = \sqrt{\frac{1}{LC} - (\frac{R}{2L})^2}$

$\therefore$ $s_1 = -\alpha + i\gamma$; $\quad s_2 = -\alpha - i\gamma$

$\therefore$ $i_c = C_1 e^{(-\alpha+i\gamma)t} + C_2 e^{(-\alpha-i\gamma)t}$

At $\quad t = 0, i_c = 0$ $\therefore$ $C_2 = -C_1$

$\therefore$ $i_c = C_1 e^{-\alpha t}.(e^{i\gamma t} - e^{-i\gamma t}) = i.2C_1 \sin\gamma t$

Since the voltage across the capacitor can not change abruptly at $t = 0^+$, it will remain zero.

Moreover, at $t = 0^+$, voltage across the resistance $i_c R$ will remain zero since at $t = 0^+$ current will remain zero.

So at $\quad t = 0^+$, voltage across the inductor $L(di_c)/dt = V$.

Differentiating the expression of i_c we get

$\frac{di_c}{dt} = i2C_1(-\alpha e^{-\alpha t}\sin\gamma t + \gamma e^{-\alpha t}\cos\gamma t)$

At $\quad t = 0^+$, $\frac{di_c}{dt} = i.2C_1.\gamma$

$C_1 = \frac{V}{i2\gamma L}$

$$\therefore \quad i_c = \frac{V}{\gamma L}e^{-\alpha t}\sin\gamma t \tag{4.48}$$

So we get an exponentially decaying sinusoidally varrying current with a frequency γ.

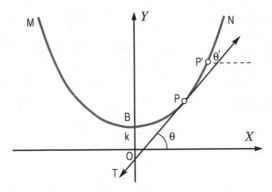

Figure 4.14: Catenary

4.13.3 Catenary

When a cable or chain is hung freely with two supports at its two ends it forms a curve called catenary (Latin catena means chain). Figure 4.14 shows a catenary curve where M and N are supporting points and B is the lowest point of the curve. Let $P(x, y)$ and $P'(x + \Delta x, y + \Delta y)$ are two points on the curve lying on the right hand side of $B(o, k)$. Let the curvilinear distance between P and P' be Δs.

Let us now isolate the arc $PP'(= \Delta s)$ from the rest of the curve. Forces associated with this small segment are as follows.

- The gravitational downward pull $\omega \Delta s$, where ω is the weight per unit length of the cable and Δs is the arc length.

- The tension T at P which makes an angle θ with x-axis.

- The tension T' at P' which makes an angle θ' with x-axis.

Let the horizontal and vertical components of T are T_x and T_y respectively. Let us now see how T_x and T_y change when we move from P to P'. The horizontal component of tension does not change during this movement. $\therefore$ $T_x = C$ where C is a constant. Since we measure the arc length s from the lowest point B and at this point $s = 0$. At this point the tension acts only in horizontal direction. Let the tension at B is denoted by T_0. $\therefore$ $C = T_0$ and $T_x = T_0$.

When we move from P to P' let the vertical component of T changes by ΔT_y and arc length changes by Δs.

$$\therefore \quad \Delta T_y = \omega.\Delta s$$

When $\Delta s \to 0$ we get $\qquad \frac{\Delta T_y}{\Delta s} = \frac{dT_y}{ds} = \omega$

Integrating the above equation we get $T_y = \omega s + C'$ where C' is the integration constant.

At B, $s = 0$ and $T_y = 0$ (only $Tx = T_0$ is present). As such, $C' = 0$

$\therefore \quad T_y = \omega s$

$\therefore \quad \frac{T_y}{T_x} = \frac{\omega s}{T_0}$

Now, $\quad \frac{T_y}{T_x} = \frac{\omega s}{T_0}$

Again,

$$\frac{dy}{dx} = \tan\theta = \frac{T_y}{T_x} = \frac{\omega s}{T_0} = \frac{s}{k} \qquad \text{where } k = \frac{T_0}{\omega}$$

$$\therefore \quad \frac{d^2 y}{dx^2} = \frac{1}{k}\cdot\frac{ds}{dx}$$

We know that,

$$\frac{ds}{dx} = \sqrt{1 + \left(\frac{dy}{dx}\right)^2}$$

$$\therefore \quad \frac{d^2 y}{dx^2} = \frac{1}{k}\cdot\sqrt{1 + \left(\frac{dy}{dx}\right)^2} \qquad (4.49)$$

Let $\quad p = \frac{dy}{dx}$

$$\therefore \quad \frac{dp}{dx} = \frac{1}{k}\sqrt{1 + p^2} \qquad (4.50)$$

$$\text{or, } \frac{1}{k}.dx = \frac{dp}{\sqrt{1 + p^2}} \qquad (4.51)$$

Integrating both sides of equation (4.51) we get

$$\frac{x}{k} = \ln\left(p + \sqrt{1 + p^2}\right) + C'' \qquad (4.52)$$

where C'' is the integration constant.

At $\quad x = 0, p = \frac{dy}{dx} = 0;$ $\quad \therefore \quad C'' = 0$

$$\therefore \quad e^{\frac{x}{k}} = p + \sqrt{1 + p^2} \qquad (4.53)$$

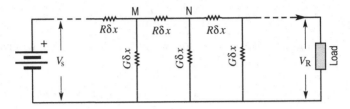

Figure 4.15: A Leaky Direct Current Line

$$\therefore \ e^{\frac{-x}{k}} = \frac{1}{e^{\frac{x}{k}}} = \frac{1}{p + \sqrt{1+p^2}} = \frac{p + \sqrt{1+p^2}}{p^2 - 1 - p^2} = \sqrt{1+p^2} - p \quad (4.54)$$

Subtracting equation (4.54) from equation (4.53) we get,

$$e^{\frac{x}{k}} - e^{\frac{-x}{k}} = 2p$$

$$\therefore \ p = \frac{dy}{dx} = \frac{e^{\frac{x}{k}} - e^{\frac{-x}{k}}}{2} = \sinh \frac{x}{k} \qquad (4.55)$$

Integrating both sides of equation (4.55) we get,
$y = k \cosh \frac{x}{k} + C'''$ where C''' is the integration constant.
At $x = 0, y = k$ and $\cosh \frac{x}{k} = 1$ $\therefore$ $C''' = 0$.

$$\therefore \ \textit{The equation of the curve is } \ y = k \cosh \frac{x}{k} \qquad (4.56)$$

4.13.4   A Leaky Direct Current Line

Let us consider a line at which a d.c. source is connected at the sending end and a load is connected at the receiving end. It is assumed that the insulation of the line is imperfect; as such, current flows through the line even when the receiving end of the line remains open. We may call this current leakage current. When a load is connected at the receiving end both the leakage current and the load current flow through the transmission line.

Let us consider a small line section MN of length dx located at a distance x from the receiving end of the transmission line (refer to Figure 4.15). Let the resistance of the transmission line be R ohm/km and the conductance of the line to ground path be G mho/km. The current entering the line section at point M is $I + dI$ and current leaving the line section is I. The voltage at m is $V + dV$ and the voltage at N is V.

We may now write the following equations for the voltage drop dV and the leakage current dI.

$$dV = R.dx.I \tag{4.57}$$

$$dI = G.dx.V \tag{4.58}$$

We can rewrite equation (4.57) as

$$I = \frac{1}{R}.\frac{dV}{dx} \tag{4.59}$$

We can also rewrite equation (4.58) as

$$V = \frac{1}{G}.\frac{dI}{dx} \tag{4.60}$$

$$\therefore \quad \frac{d^2V}{dx^2} = R.\frac{dI}{dx} = RGV = \alpha^2 V \tag{4.61}$$

$$\text{where} \quad \alpha = \sqrt{RG}$$

$$\text{Similarly,} \quad \frac{d^2I}{dx^2} = G.\frac{dV}{dx} = RGI = \alpha^2 I \tag{4.62}$$

Now we can consider the following equations as the solutions of equation (4.61) and equation (4.62).

$$V = A.e^{\alpha x} + B.e^{-\alpha x} \tag{4.63}$$

In the above equation A and B are constants.

$$I = C.e^{\alpha x} + D.e^{-\alpha x} \tag{4.64}$$

In equation (4.64) C and D are constants.

$$\text{At } x = 0, \quad I = I_R \text{ and } V = V_R \tag{4.65}$$

$$\therefore \quad V_R = A + B \text{ and } I_R = C + D$$

We can rewrite equation (4.64) as follows,

$$
\begin{aligned}
I &= (I_R - D).e^{\alpha x} + D.e^{-\alpha x} \\
&= I_R.e^{\alpha x} - 2D.\frac{(e^{\alpha x} - e^{-\alpha x})}{2} \\
&= I_R.\frac{e^{\alpha x}}{2} + I_R.\frac{e^{\alpha x}}{2} + I_R.\frac{e^{-\alpha x}}{2} - I_R.\frac{e^{-\alpha x}}{2} - 2D\sinh\alpha x \\
&= I_R.\frac{(e^{\alpha x} + e^{-\alpha x})}{2} + I_R.\frac{(e^{\alpha x} - e^{-\alpha x})}{2} - 2D\sinh\alpha x \\
&= I_R\cosh\alpha x + I_R\sinh\alpha x - 2D\sinh\alpha x
\end{aligned}
$$

$$
\therefore \quad I = I_R\cosh\alpha x + (I_R - 2D)\sinh\alpha x \qquad (4.66)
$$

We can rewrite equation (4.63) as follows,

$$
\begin{aligned}
V &= (V_R - B).e^{\alpha x} + B.e^{-\alpha x} \\
&= V_R.e^{\alpha x} - 2B.\frac{(e^{\alpha x} - e^{-\alpha x})}{2} \\
&= V_R.\frac{e^{\alpha x}}{2} + V_R.\frac{e^{\alpha x}}{2} + V_R.\frac{e^{-\alpha x}}{2} - V_R.\frac{e^{-\alpha x}}{2} - 2B\sinh\alpha x \\
&= V_R.\frac{(e^{\alpha x} + e^{-\alpha x})}{2} + V_R.\frac{(e^{\alpha x} - e^{-\alpha x})}{2} - 2B\sinh\alpha x \\
&= V_R\cosh\alpha x + V_R\sinh\alpha x - 2B\sinh\alpha x \\
\therefore \quad V &= V_R\cosh\alpha x + (V_R - 2B)\sinh\alpha x \qquad (4.67)
\end{aligned}
$$

From equation (4.60)

$$
V = \frac{1}{G}.\frac{dI}{dx} = \frac{1}{G}.[I_R\alpha\sinh\alpha x + (I_R - 2D)\cosh\alpha x] \qquad (4.68)
$$

Comparing coefficients of $\sinh\alpha x$ of equation (4.67) and equation (4.68) we get,

$$
(V_R - 2B) = \frac{I_R}{G}\alpha = \frac{I_R}{G}.\sqrt{RG} = I_R.\sqrt{\frac{R}{G}} = I_R.\gamma \qquad (4.69)
$$

$$
\text{where } \gamma = \sqrt{\frac{R}{G}}
$$

So we may now rewrite equation (4.67) as follows

$$V = V_R \cosh \alpha x + I_R . \gamma \sinh \alpha x \qquad (4.70)$$

The above equation shows how voltage at a point in the line changes with change in x.

Comparing the coefficients of $\cosh \alpha x$ of equation (4.67) and equation (4.68) we get,

$$\frac{I_R - 2D}{G} . \alpha = V_R$$

$$\therefore \quad I_R - 2D = V_R . \frac{G}{\alpha} = \frac{V_R .}{\gamma}$$

We may now rewrite equation (4.66) as

$$I = I_R \cosh \alpha x + \frac{V_R}{\gamma} \sinh \alpha x \qquad (4.71)$$

Equation (4.71) shows how current through the line changes with x.

4.13.5 Dynamics of moving bodies

Displacements of two separate bodies moving along a co-ordinate line are given below as functions of time.

$$x_1 = a . \cos kt + b . \sin kt$$

$$x_2 = c . \cosh kt + d . \sinh kt$$

So the velocities v_1, v_2 and accelerations a_1, a_2 of these two bodies will be as follows.

$$v_1 = \frac{dx_1}{dt} = -ak . \sin kt + bk . \cos kt$$

$$a_1 = \frac{dv_1}{dt} = -ak^2 . \cos kt - bk^2 . \sin kt = -k^2 . x_1$$

$$v_2 = \frac{dx_2}{dt} = ck . \sinh kt + dk . \cosh kt$$

$$a_2 = \frac{dv_2}{dt} = ck^2 . \cosh kt + dk^2 . \sinh kt = k^2 . x_2$$

It is important to note here that

$$\frac{d \sinh x}{dx} = \cosh x \quad \text{and} \quad \frac{d \cosh x}{dx} = \sinh x.$$

So for the first body the acceleration is proportional to displacement and is directed towards origin (since the sign is negative). The displacement is expressed in terms of circular functions $\cos kt$ and $\sin kt$ which are periodic functions of time t. So the movement of the first body is oscillatory in nature.

We observe that for the second body also the acceleration is proportional to displacement but is directed away from the origin (since the sign is positive). The displacement is expressed in terms of hyperbolic functions $\cosh kt$ and $\sinh kt$ which keep on increasing with time t. So the second body keeps on drifting monotonically with time without any bounds.

4.14   Graphs of different hyperbolic functions

In Figure 4.16 graphs of hyperbolic functions $\sinh x$, $\cosh x$, $\tanh x$, $\coth x$ etc. have been shown.

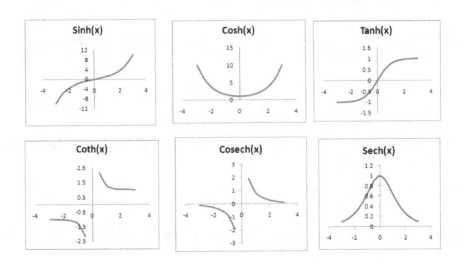

Figure 4.16: Graphs of different hyperbolic functions

4.15 Are the hyperbolic functions periodic?

It is well known that circular functions like $\sin\theta$, $\cos\theta$ are periodic functions. Their graphs also support this idea. On the other hand, graphs of hyperbolic functions like $\sinh\alpha$, $\cosh\alpha$ (refer Fig. no. 4.16) may give us an impression that these are aperiodic. It is important to note that critical examination of these functions reveal that they are also periodic.

We know that $\cosh\alpha = \frac{e^\alpha + e^{-\alpha}}{2}$ [refer equation (4.25)]

Again we know that $\cos\theta = \frac{e^{i\theta} + e^{-i\theta}}{2}$

[refer equation (2.34)]

If $\theta = i\alpha$,

$$\cos\theta = \cos i\alpha = \frac{e^{i(i\alpha)} + e^{-i(i\alpha)}}{2} = \frac{e^\alpha + e^{-\alpha}}{2} \qquad (4.72)$$

So we find that cosine hyperbolic of argument α and cosine circular of argument $i\alpha$ are same. As such, $\cosh\alpha$ may have a periodicity of $i.2\pi$.

We know that $e^{i\alpha} = \cos\alpha + i\sin\alpha$

$$\therefore\ e^{i2\pi} = \cos 2\pi + i\sin 2\pi = 1 + i0 = 1$$

$$\text{and } e^{-i2\pi} = \cos(-2\pi) + i\sin(-2\pi) = 1 + i0 = 1$$

$$\therefore\ e^{(\alpha+2\pi i)} = e^\alpha . e^{2\pi i} = e^\alpha$$

$$\therefore\ e^{(-\alpha-2\pi i)} = e^{-\alpha} . e^{-2\pi i} = e^{-\alpha}$$

$$\therefore\ \cosh(\alpha + 2\pi i) = \frac{e^{\alpha+2\pi i} + e^{-\alpha-2\pi i}}{2} = \frac{e^\alpha + e^{-\alpha}}{2} = \cosh\alpha$$

Similarly, $\sinh(\alpha + 2\pi i) = \frac{e^{\alpha+2\pi i} - e^{-\alpha-2\pi i}}{2} = \frac{e^\alpha - e^{-\alpha}}{2} = \sinh\alpha$

We can now generalise the idea by writing the following two equations.

$$\cosh(\alpha + 2n\pi i) = \cosh\alpha$$

$$\text{and } \sinh(\alpha + 2n\pi i) = \sinh\alpha$$
$$\text{where } n = 1, 2, 3, \cdots$$

So we note that cosine hyperbolic and sine hyperbolic functions repeat themselves when their arguments are increased by integral multiple of 2π. Unlike circular functions $\sin\theta$ and $\cos\theta$, *periodicity of hyperbolic cosine and sine functions can not be observed visually*

*from their graphs drawn on a two dimensional plane. To visualize it
we need to construct a number of planes parallel to the real $X - Y$
plane where these planes are separated by a distance 2π along the Z
direction which indicates the imaginary axis.*

Let us now examine the periodicity of hyperbolic tan function.

We know that $e^{i\pi} = \cos \pi + i \sin \pi = -1 + i0 = -1$

$$\therefore \quad e^{i\pi + \alpha} = e^{i\pi}.e^{\alpha} = -e^{\alpha}$$

Since $e^{-i\pi} = \cos(-\pi) + i \sin(-\pi) = -1 + i0 = -1$

$$\therefore \quad e^{-i\pi - \alpha} = e^{-i\pi}.e^{-\alpha} = -e^{-\alpha}$$

$$\therefore \quad \tanh \alpha \;=\; \frac{e^{\alpha} - e^{-\alpha}}{e^{\alpha} + e^{-\alpha}} = \frac{-e^{\alpha + \pi i} + e^{-\alpha - \pi i}}{-e^{\alpha + \pi i} - e^{-\alpha - \pi i}}$$

$$=\; \frac{e^{\alpha + \pi i} - e^{-\alpha - \pi i}}{e^{\alpha + \pi i} + e^{-\alpha - \pi i}} = \tanh(\alpha + \pi i)$$

So we find that tanh function has a period of πi.

4.16   Expressions for inverse hyperbolic functions

Before we deal with inverse hyperbolic functions we need to derive an
important identity.

We know that $\cosh \alpha = \frac{e^{\alpha} + e^{-\alpha}}{2}$ and $\sinh \alpha = \frac{e^{\alpha} - e^{-\alpha}}{2}$

$$\therefore \quad \cosh^2 \alpha - \sinh^2 \alpha = \left(\frac{e^{\alpha} + e^{-\alpha}}{2} \right)^2 - \left(\frac{e^{\alpha} - e^{-\alpha}}{2} \right)^2 = 1$$

Let us consider the hyperbola $x^2 - y^2 = 1$

Let $x = \cosh \alpha \quad \therefore \quad \sinh \alpha = \sqrt{\cosh^2 \alpha - 1} = \sqrt{x^2 - 1}$

But we know that $\cosh \alpha + \sinh \alpha = e^{\alpha}$

$$\therefore \quad x + \sqrt{x^2 - 1} = e^{\alpha}$$

$$\therefore \quad \alpha = \cosh^{-1} x = \ln\left(x + \sqrt{x^2 - 1} \right) \tag{4.73}$$

Now let $x = \sinh \alpha$ $\therefore$ $\cosh \alpha = \sqrt{\sinh^2 \alpha + 1} = \sqrt{x^2 + 1}$

$$\therefore \quad x + \sqrt{x^2 + 1} = \cosh \alpha + \sinh \alpha = e^\alpha$$

$$\therefore \quad \alpha = \sinh^{-1} x = \ln \left(x + \sqrt{x^2 + 1} \right) \qquad (4.74)$$

Let $x = \tanh \alpha = \frac{e^\alpha - e^{-\alpha}}{e^\alpha + e^{-\alpha}}$

$\therefore \quad \frac{1+x}{1-x} = \frac{e^\alpha}{e^{-\alpha}} = e^{2\alpha}$

$$\therefore \quad \alpha = \tanh^{-1} x = \frac{1}{2}.\ln \frac{1+x}{1-x} \qquad (4.75)$$

$$\text{sech}^{-1} x = \cosh^{-1} \frac{1}{x} = \ln \frac{1 + \sqrt{1 - x^2}}{x} \qquad (4.76)$$

$$\text{cosech}^{-1} x = \sinh^{-1} \frac{1}{x} = \ln \frac{1 + \sqrt{1 + x^2}}{x} \qquad (4.77)$$

$$\coth^{-1} x = \tanh^{-1} \frac{1}{x} = \frac{1}{2}.\ln \frac{1-x}{1+x} \qquad (4.78)$$

4.17 Infinite series representation of cosh x and sinh x

In Chapter 2 (refer equation (2.14)) the following has been derived.

$$e^x = 1 + x + \frac{x^2}{2!} + \frac{x^3}{3!} + \frac{x^4}{4!} + \frac{x^5}{5!} + \cdots \infty$$

$$\therefore \quad e^{-x} = 1 - x + \frac{x^2}{2!} - \frac{x^3}{3!} + \frac{x^4}{4!} - \frac{x^5}{5!} + \cdots \infty$$

$$\therefore \quad \cosh x = \frac{e^x + e^{-x}}{2} = 1 + \frac{x^2}{2!} + \frac{x^4}{4!} + \cdots \infty \qquad (4.79)$$

$$\therefore \quad \sinh x = \frac{e^x - e^{-x}}{2} = x + \frac{x^3}{3!} + \frac{x^5}{5!} + \cdots \infty \qquad (4.80)$$

If we put $x = i\theta$ in equation (4.79) we get the infinite series for the circular function $\cos \theta$.

$$\cosh i\theta = 1 - \frac{\theta^2}{2!} + \frac{\theta^4}{4!} - \cdots \infty = \cos \theta$$

Similarly we put $x = i\theta$ in equation (4.80) we get i times the infinite series for circular function $\sin \theta$

$$\sinh i\theta = i\theta - i\frac{\theta^3}{3!} + i\frac{\theta^5}{5!} - \cdots \infty = i \sin \theta$$

$$\sin \theta = \frac{\sinh i\theta}{i} = -i \sinh i\theta$$

So we once again observe that circular angle and hyperbolic angles are in quadrature.

4.18   Historical Development of the concept of hyperbolic functions

- Gerhard Kremer (1512- 1594) (Latin Name Gerhardus Mercator) developed his Mercator's projection map in 1592 in which he used a hyperbolic function without explicitly mentioning it (afterwards mathematicians identified it). His work made a break through in navigation.

- In fifteenth century da Vinci was interested in knowing the shape of the curve assumed by a flexible, extensible cord hung freely from two fixed points. In 1690 Christian Huygens wrote a letter to Leibniz in which he termed it as Catenary. Afterwards in 1691 Huygens, Leibniz and Johann Bernoulli provided independent correct solutions to this problem. None of them, however, associated his solution with hyperbola. Now we express it as $y = k \cosh \frac{x}{k}$

- Euler, in the Volume 1 of his *Introductio in Analysin Infinitorum* (1745) used the expressions $\frac{e^x + e^{-x}}{2}$ and $\frac{e^x - e^{-x}}{2}$ while deriving infinite product representations for the sine and cosine functions. He also did not associate these with hyperbola.

- Vincenzo Riccati (1707 -1775), second son of renowned mathematician Jacopo Riccati (famous for Riccati equation in differential equations), introduced the use of hyperbolic functions while working with cubic equations. In his two-volume *Opuscula ad res physicas et mathematicaps ertinentium* (1757-1762) he defined sine hyperbolic and cosine hyperbolic functions and used the notations Sh ϕ and Ch ϕ respectively (for circular sine and cosine functions he used the notations Sc ϕ and Cc ϕ respectively).

He further developed addition formulas and other identities for hyperbolic functions and their derivatives in his book *Institutiones analyticae*(1765-1767) written in collaboration with Girolamo Saldini. In this book he also expressed hyperbolic functions in terms of exponential functions.

- Johann Heinrich Lambert (1728 -1777) first introduced hyperbolic functions in trigonometry in a paper published in 1768. He also associated these with geometry. He parameterised an equilateral hyperbola $x^2 - y^2 = 1$ by $\cosh \alpha$ and $\sinh \alpha$. He called α -the transcendental angle and proved that it is twice the area of the hyperbolic sector. He also used hyperbolic functions while solving certain types of differential equations.

- De Foncenex, a student of Lagrange, interestingly pointed out that the circular sector and hyperbolic sector that correspond to the same abscissa bear a ratio $1 : \sqrt{-1}$ i.e., $1 : i$. It means that if ordinate on a circle is considered as a real quantity then ordinate on a hyperbola needs to be shown as an imaginary quantity. We may, however, show ordinate on a hyperbola as a real quantity; in that case ordinate on a circle is considered as an imaginary quantity.

- Christoph Gudermann(1798 — 1852) published a paper in 1832 in which he investigated on the solution of Mercators Projection. His also derived expressions relating hyperbolic angles with circular angles.

- T. H. Blakesley used hyperbolic angles in a paper entitled *Alternating Currents of Electricity* published in 1889 in London.

- Sir J. J. Thomson used hyperbolic angles in a paper entitled *On the Heat Produced by Eddy Currents in an Iron Plate exposed to an Alternating Magnetic Field* published in The Electrician, 1891, Vol. XXVIII, p. 599

- A. E. Kennelly advocated the use of hyperbolic functions in the study of transmission line through numerous papers. He also published an important book *Hyperbolic functions applied to Electrical Engineering* in 1916.

Bibliography

1. Boyajian, A., *Physical Interpretation of Complex Angles and Their Functions*, Jour., A.I.E.E., February, 1923. Pp. 155 — 164

2. Kennelly, A. E., Discussion on *Physical Interpretation of Complex Angles and Their Functions*, Jour., A.I.E.E., February, 1923.

3. Loew, E. A., 1928 *Electrical Power Transmission* First Edition, McGraw Hill Book Co.

4. Fehr, Howard F., *Mathematics for Teachers* D. C. Heath AND Company, Boston

5. Weinbach, M, P., 1924 *Principle of transmission in Telephony* The Macmillan Company, New York

6. Kennelly, A. E., 1912 *The Application of Hyperbolic Functions to Electrical Engineering Problems* University of London Press, London

7. Kennelly, A. E., *The Present Status Of Complex Angles In Their Applications To Electrical Engineering* Electrical Society Journal (Japan), Feb, 1942 pp 166-185 (Re-published)

8. Barnett, J. H., *Enter Stage Center: The Early Drama of the Hyperbolic Functions* Mathematics Magazine, Vol. 77, No. 1, Feb, 2004, pp. 15 -30

9. Burn, B., *Gregory of St. Vincent and the Rectangular Hyperbola* The Mathematical Gazette, Vol. 84, No. 501., Nov., 2000, pp. 480-485.

10. Mcmahon, J.,1906 *Hyperbolic Functions* 4th Edition Enlarged, John Wiley & Sons,New York, Chapman & Hall, London.

11. Markushevich, A. I., 1966 *The Remarkable sine functions* American Elsevier Publishing Company Incorporated, New York

12. Maor E., 1998 *Trigonometric Delight* Princeton University Press, New Jersey

13. Kline M. *Calculus- An Intuitive and Physical Approach* 2nd Edition, Dover Publication

14. Boyajian, A. *Origin and Meaning of Circular and Hyperbolic Functions in Electrical Engineering* Journal of Franklin Institute, Vol. 249, Issue 2, February 1950, pp. 117-131.

15. Shervatov, V. G. 2007 *Hyperbolic Functions* Dover Publications Inc., Mineola, New York

Index

Printed in the United States
by Baker & Taylor Publisher Services